Mr. 22nd Infantry

Mr. 22nd Infantry

Michael D. Belis

Deeds Publishing | Athens

Published by Deeds Publishing in Athens, GA
www.deedspublishing.com

Printed in The United States of America

Cover and interior design by Deeds Publishing

ISBN 9978-1-961505-21-6

Books are available in quantity for promotional or premium use. For information, email info@deedspublishing.com.

First Edition, 2024

10 9 8 7 6 5 4 3 2 1

Dedicated to John King

Be at rest dear friend,
Your work on this earth is done ... and done well.

Contents

Acknowledgements

Thanks to Dr. John R. King Jr., the nephew of Earl W. "Lum" Edwards. This story would not have been told except for John. Over the course of several years, and hundreds of phone conversations, I came to see John as a good friend. His zeal and persistence, to getting the service of his Uncle "Lum" assembled and set down for posterity was inspiring.

John had gathered photos and documents of his uncle's service, but had contracted MS, and was thus prevented from writing the story himself. Karen Scott, the cousin of John Dowdy, who was a fellow officer in the 22nd Infantry with Lum Edwards, knew John, and got John and me talking directly to each other. John was gently persuasive, and in short order, I agreed to tackle the job.

My thanks go out to Karen. Little did I know at the time that she had introduced me to someone I would come to admire and view as a friend. Karen is dedicated to preserving the history of her cousin, Lieutenant Colonel John Dowdy, and the 22nd Infantry in WW2. She did a remarkable job of gathering research on Dowdy and seeing that his story was told. And then she was instrumental in helping to get the story of Lum Edwards told.

In concert with John King, I did my own research then, and gathered material that was different from that which John had assembled. I ended up with a quantity that was about as much material as John had. Putting his research and my research together, and fashioning the story, took several years. John and I would discuss it frequently as I would send him pages when I completed them.

Sadly, John passed away in early 2022. He never got to see this published version of his uncle's story. In 2017, I did print out several hard copies of the finished manuscript and sent one to John and another to his mother Gladys (sister of Earl Edwards) and also sent John many digital copies as well. He was delighted with the final result, and I felt satisfied in securing his approval of the completed work.

Thanks to John's mother, Gladys King. She did all the physical work involved in packaging and sending me hard copies of John's research. She sent me by e-mail, and through the post office mail, copies of all the photos John had accumulated. Gladys is a real trooper and was instrumental to getting the Lum Edwards' story told.

Thanks to Bill Andrews for allowing me to use quotes from his book on his father's service with the 22nd Infantry "*The Road from Caledonia to Canisy, One man's Journey from Home Through World War II And Back*". Bill's book contains excellent descriptions of the action seen by the Regiment during June and July 1944. Bill's research and writing are outstanding, and his book is an excellent tribute to his father.

Finally, thanks to Bob Babcock, Jan, Mark, Matt, and the staff at Deeds Publishing. Bob is dedicated to telling the stories of our veterans, and especially passionate in preserving the history of the 4th Infantry Division and the 22nd Infantry Regiment. It was at

Bob's urging that this published version of the story of Lum Edwards has finally come to be.

        —**Steadfast and Loyal! Deeds Not Words!**

Preface

Many years after his retirement from the Army, Lum Edwards wrote his memoirs. His narrative is used throughout this presentation. He began his memoirs with his Army service so the story of his childhood and early life is told in the following pages by his younger sister, Gladys. Lum wrote about his early military service and his time with the 22nd Infantry up to the D-Day invasion. He did not write anything about his wartime service in Europe. His reason for doing that was written in his memoirs with the following: "Since this story can be just so long, I will skip combat. That would be a book, a very long book all by itself."

What a sad decision on his part! It would have been fascinating to have a record of his days in combat as seen through his own eyes. We wish Lum had indeed written that book on his combat service. We would have loved to read it.

Don Warner began his service in the 22nd Infantry as a Platoon Leader, rose to Company Commander, then Assistant Regimental S-3 Operations Officer under Lum Edwards, and finally ended his service with the Regiment as S-3 Operations Officer for 3rd Battalion. Warner related that Lum was given the title of "Mr. 22nd Infantry" by Colonel John Ruggles, who commanded the 22nd Infantry Regiment in 1945-1946. Warner says the title was conferred upon him by Ruggles because Lum was the longest serving member of the Regiment. The case could be made that Thomas Kenan actually served longer in the Regiment than Lum did…by nearly one whole year, though Kenan did serve with 4th Infantry Division staff for a few months during the war before coming back to the 22nd Infantry Regiment.

However, it was Lum Edwards and Chaplain Bill Boice who were the driving force behind the creation of the 22nd Infantry Regiment Association in the late 1940's. And it was Lum who guided the Association through its early years and kept the comradery of its members alive. And that organization (renamed the 22nd Infantry Regiment Society) has continued for almost eighty years in carrying on the traditions of the Regiment and preserving its history. In that sense, the title of "Mr. 22nd Infantry" is well deserved by Earl W. "Lum" Edwards.

In a letter to Don Warner many years after the war, Major General Charles T. Lanham who, as a Colonel, commanded the 22nd Infantry Regiment from July 1944 to March 1945 wrote:

"The 22nd was a magnificent regiment with magnificent battalion and company commanders. Johnny Ruggles, my exec, was second to none in that capacity and I have never seen Lum Edwards' equal as an S-3. We had a battle team that could do anything, including miracles, and did. It was proud of itself, still is, and darn good reason to be. I doubt that this country will ever see its equal again."

—Michael Belis
DMOR 22nd Infantry Regiment
SGT C 1/22 Infantry 1970-1971
Carencro, Louisiana
July 7, 2017

1. The Early Years

Earl William Edwards was the son of Homer Dee Edwards Sr., and Sara Lou "Lula" Wilks. His father was from Mississippi and had been working in Texas where he met his future wife. He moved back to Mississippi to work for the railroad. Lula joined him and they were married in Tunica County, Mississippi. Earl Edwards was born in Banks, Mississippi on September 3, 1917.

Sometime after Earl was born, the family moved to Cruger, Mississippi. Earl's sister Gladys describes their early life:

"Mother and Daddy lived in what the railroad called Camp cars. They were several special rail cars that were outfitted for living accommodations. They lived in them for some time until the section house was built. They were called section houses as Daddy was a section foreman. He had responsibility of a certain section of the track.

There were six children in the family. Earl William Edwards, Jesse Ray Edwards, Mattie Charline Edwards, Gladys Nell Edwards, Fred Fields Edwards, and Homer Dee Edwards Jr. Ray was tragically killed by a train. He was the child born between Earl and Charline. Mother was delivering clothes to a lady to be ironed. She had Earl, Ray, and a friend in the car with her. She did not see or hear the train at the crossing as it was a blind crossing at that time.

Her car stalled on the track and the train hit it. Ray was killed and none of the others were hurt.

In the depression of 1929, Daddy was reduced to being a traveling section foreman. He did not want to do that, so he quit working for the railroad. He had owned a small grocery store before that but had a Negro man running it for him. When he left the railroad, Daddy took over running the store. It was not located in the main stream of stores in Cruger at the time and Daddy thought he could make more money if we moved up the street. (Like many small towns, Cruger had a single street where most of its businesses were located.) This we did and that is where we were when that picture of Earl and Mr. Vowell was taken. Earl was working there to help out so mother could stay home and run the house and take care of the rest of the kids.

Earl worked in his father's store, waited on customers, and swept up the place.

Gladys wrote: "Bananas were purchased by the stalk and were displayed by hanging them. As a rule, most were sold by the single banana. His father told him not to help himself to those still on the stalk but that he could eat all that became over ripe and fell on the floor. Lum mentioned that such a banana was, in his opinion, 'vintage' and therefore belonged to him."

Earl, as I knew him, was never athletically inclined. He never played any sports that I know of. In fact, the high school he attended was so small that there were only two members in his graduating class. No wonder he did not play on any teams. Earl liked to read a lot in his early years and that continued until the time of his death. Daddy used to tell him to quit reading and get off the porch and go play ball or some other activity.

We all belonged to the Southern Methodist Church as it was known then.

In Earl's teenage years, he was lucky enough to get a job working on the construction of Highway 49. He was on the street of Cruger one day doing nothing and a man approached him and asked him if he could drive a truck. Of course he said yes, and the man said go on over to the highway and you will be given a truck hauling dirt for the building of the road bed. The work was long and hard. The hours were from 8 a.m. until 6 p.m. with no stopping. They were not allowed to stop for lunch. Every day, Mother would fix his lunch and put it in a card board box with a quart fruit jar filled with iced tea. One of us would go over to the highway and sit on the road side until he came by, pulled off the way just enough for us to hand him the lunch in the window of the truck, and he would pull back on the road and keep going, eating as he drove along. I remember doing that many times for him.

This is one way Earl got some of the money to start classes at Mississippi State. We did not have a car at the time and Daddy arranged with another family that was taking their son over to take Earl also.

As young children, we had no indoor bath facilities, and our Saturday night bath was in a number 3 tin tub in the kitchen. Water was heated on the wood stove and poured in the tub till it was filled enough to get a good bath.

When we were small, we all liked the leg and thigh of fried chicken. Earl used to try to trick Charline and me out of wanting them by telling us if we ate the back and wings they would make us pretty. That worked for a while, but soon it became a race who could get the leg and thigh first.

I remember one time Mother and Daddy had a serious argument. Whatever it was about, Earl, about four years old at the time, took Mother's side and forever became her favorite child. It never bothered any of us. We all freely admitted it and talked and laughed about it. In fact he proved to be it in every sense of the word throughout his life. If it were not for him and his generosity, none of the rest of us would have had a chance for college. Only Fred and I took full advantage of it and finished college.

As far as Earl going to Miss. State, it was a forgone conclusion that was where he would go if he could. Ole Miss was much more expensive and known as a society type of school that was mostly concerned with turning out lawyers and doctors. Earl wanted to be neither. He had wanted to come back to Cruger and buy a farm, but that was not to be, due to many circumstances."[1]

Earl Edwards graduated as Class President of Cruger High School in 1935 and enrolled in Mississippi State University to pursue a degree in Agricultural Engineering. His plan was to get a degree and return home to farm. At that school, Military Science was a required subject for physically fit males for the first two years of their degree programs. He found that he liked the military courses and consequently made his best grades in them. He continued his Military Science curriculum in his junior and senior years, enrolling in Advanced Military Science which also paid $9.00 per month. That helped financially, as the Great Depression of the 1930's was still affecting the country.

While at Mississippi State, he was given the nickname of "Lum" by one of his classmates. A popular radio show of the time was called "Lum and Abner" and it concerned the lives of people in a small town in Arkansas. The main characters of the show ran a country general store and one of them went by the name of Lum Edwards. In keeping with the "country" theme of the show, on radio the name was pronounced "Eddards." The similarities were too much for the classmate — the last name of Edwards, the small "country" town origin, and the fact that Edwards had worked in a general store. So he began calling Earl Edwards "Lum". The nickname stuck with him for the rest of his life.

In his senior year at Mississippi State, Edwards made the list of "Who's Who in American Universities and Colleges." He was President and Program Chairman of the A.E.A.E., a member of Scabbard and Blade and on the Executive Committee of the Agriculture Club.

Edwards was also at the top of his military class and was appointed Lieutenant Colonel (second in Command) of the school's Cadet Corps. He would have been Cadet Colonel (first in Command of the Cadet Corps) but the year he was a senior the Cadet Colonel position was appointed from the Artillery branch of Cadets and Edwards was assigned to the Infantry branch. The school policy was for the Infantry and Artillery branches to alternate Command of the Cadet Corps each successive year, and his senior year happened to fall upon a year in which the Artillery branch was to be in Command.

After completing four years of ROTC, graduates would receive a commission in the Army Reserve Forces. Being the top academic student in the ROTC program, as graduation approached, Edwards was invited to apply for a commission in the Regular Army. In 1939, the farming industry in the South was in bad shape. Future prospects in Edwards' chosen profession did not look good. A career in the Army however held more promise. After twenty years of service, it would be possible to retire with a monthly income and he would still be young enough to pursue other interests or livelihoods.

Edwards therefore applied for the commission, appeared before a selection board, and was informed that he had been selected to receive the commission. He was to await orders to report at a future time. At the same time, he received a letter from the U.S.

Marine Corps offering him a permanent commission providing he passed their physical exam. George Gober, a good friend of his all through ROTC was named as alternate should Edwards not take the commission. Edwards thought it would be a good idea to take the exam as insurance against something going wrong with his Army commission. Edwards and Gober went together to Pensacola for the physical exam. Edwards failed the color blind test, was rejected, and Gober got the commission.

Immediately after the exam, Edwards and Gober reported to Fort McClellan, Alabama to serve two weeks active duty as Reserve Officers as a mandatory requirement upon their graduation from ROTC. The official date of Edwards' commission as an officer in the Army Reserve Forces was May 22, 1939. This first period of active duty for Edwards was from June 11 to June 24, 1939. After returning home from active duty, Edwards learned that a graduate of another university had been given the Regular Army commission he had been promised. An important Senator had intervened and caused the commission to be given to a personal favorite.

With his future in the military now eliminated, Edwards had to take a job in Tennessee with an electrical construction company building power lines to bring electricity to the rural areas of that state. It was grueling work, cutting and blasting right of ways, erecting poles, stringing power lines, hooking up to houses. He wore climbing spikes and climbed trees to cut down branches and he wondered how a college graduate like himself had come to be in such a situation.

His position as top graduate in ROTC from Mississippi State had not failed him. He received a letter from the Army proposing that he go on active duty for a six month tour as a Reserve officer, during which time he could compete for a commission in the Regular Army. He took a leave of absence from his job and reported for duty on December 1, 1939 with the 11th Infantry Regiment, temporarily posted at Fort McClellan, Alabama. The 11th Infantry was quartered in tents in an area off the main post. This was the Army before the draft began. Officers and NCO's were career men, many having been in the Army for 15-20 years. Most enlisted men came from poor backgrounds and had little education. A great many enlisted because jobs during those hard economic times of the 1930's were in short supply. Many men had gotten in trouble with the law and enlisted to avoid prison time. It was even customary in those days for a judge to let a young offender escape jail or prison time and therefore gaining a criminal record, if the offense was not too serious and the offender would instead enlist in one of the military services. It was believed that such an offender would benefit from a life of military discipline and the experience would straighten out his life and allow him to become a productive member of society.

At McClellan, the 11th Infantry spent their time doing close order drill, inspections, field marches, and training exercises. Edwards recalled those first days he spent as an officer:

"The non-commissioned officers were men with 15 to 30 years of service and real professional soldiers. They had joined the Army as young men, liked the life, and worked their way up the ladder the hard way. They knew their business. I learned very quickly that you could depend on them for help and learn a lot by watching how they carried out their duties. They earned their place as leaders by being the best and toughest soldiers in their unit with a willingness to accept command responsibility. If an enlisted man did not do his job, he had a real man to answer to. Some were hard-living, and perhaps hard-drinking off the post, but when on duty they were all business. I had great respect for all I knew. They were indeed the backbone of the Army.

"Our days were spent doing close order drill, learning how to handle our weapons, going on field marches and exercises. It was fun and I enjoyed all of it. Being a very junior officer and with very few of us available at that, I quickly caught a week-duty to mount and supervise guard duty. As a matter of training we had a contingent of guards posted about camp under the supervision of a sergeant of the guard and Officer of the Day. To begin their tour of duty, about 25 to 30 men were assembled on the parade ground to be inspected and instructed by the Officer of the Day. This meant that I had to get up in front and issue a standard set of orders and instructions. I thought I had memorized the words but possibly, because of stage fright, and just possibly a lack of preparation, I found myself mute. What an embarrassing moment!! I couldn't think of a word to say. All the men stood at attention awaiting my instructions. The Sergeant of the Guard was a red-headed and freckle-faced Irishman and a grizzled old veteran of many years of service. His position was standing just to the front and right of my position. He quickly sensed my problem and without moving his head to give away what he was doing he whispered what I should say out of the side of his mouth. So, with his help, I was able to get through my first Army crisis in good shape. I was forever grateful to him, and it taught me a valuable lesson — how important it was to have the respect and loyal support of all the people under your command and supervision. Many times in life this can make or break you." [2]

Edwards recalled another incident that illustrated life in the Army at that time:

"This reminds me of one incident that early on showed me how some old Army sergeants commanded their platoons. We were on an overnight march and made camp late one evening in the snow. Our pup tents were pitched in a long row, I was standing with the Platoon Sergeant at the head of the street watching the men make ready for the night. Someone had arranged for a hay wagon to drop off some bales of hay at the foot of the street so the men could put some in their tents for warmth. As I looked down the street, I saw several of the men arguing rather vociferously over the division of the hay. I pointed this out to the Sergeant and said it looked as if it might develop into a fight. The Sergeant indicated he didn't think it was something to worry about. But after a bit, it seemed to get worse, and I was somewhat worried that a fight might develop on my watch right in my sight. So I indicated to the Sergeant that I was quite concerned. He saw I wasn't going to give it up, so he casually walked down the street and stood a minute or two talking to the two most belligerent men and suddenly knocked one of them down with his fist. Then he strolled casually back up the street, stopping to chat with men on the way. When he got back to me he said, "Everything is alright now, Lieutenant." This is how I learned the way some sergeants in that day and time maintained discipline." [3]

During the cold winter of 1939-1940, Edwards came down with strep throat and spent a week in the hospital. In the spring of 1940, the 11th Infantry moved to Fort Benning, Georgia. Almost immediately they were assigned to take part in the Louisiana maneuvers for that year. The site of the maneuvers was in the immense Kisatchie National Forest of Louisiana, located north of Alexandria. The 11th Infantry had to travel by trucks to get there. Their route took them from Fort Benning to Mobile, Alabama, then along the Mississippi Gulf Coast to New Orleans and finally up to Dry Prong, Louisiana. Edwards recalled the journey:

"You would not believe the ancient, battered old trucks we made the trip in. They seemed to use more oil than gas and kept breaking down all the time. The Army was in real bad shape in those early days in terms of equipment of all kinds…We would make several hundred miles a day and then camp for the night. We camped one night just outside of New Orleans. A group of us went into town and toured the city. I thought I was really seeing the world…Arrival in Dry Prong, or rather in the piney woods near Dry Prong, brought on hot, dry weather. We quickly began a training program that had us maneuvering and training night and day as the Army desperately tried to get us ready for war. Ticks were so bad that each man had to be paired with a 'Tick Buddy' as they called it. Every two hours by the clock we had to stop and shed our clothes so your Tick Buddy could clean off the ticks on you." [4]

While in Louisiana, Edwards appeared before a Board of Officers to be considered for an appointment in the Regular Army. The Board President explained what had happened to Edwards' Honor Graduate Appointment. Edwards had been the number one choice and had been selected to receive the commission, but an important (though unnamed) Senator had intervened and caused Edwards' selection to be dropped in favor of one Morgan Roseborough from Greenwood, Mississippi. Roseborough was a graduate of Ole Miss. The Board President went on to express his disgust with this turn of events and assured Edwards his commission would be upheld this time around. [5]

At the end of the maneuvers, the 11th Infantry returned by train to their home base at Fort Benjamin Harrison, Indianapolis, Indiana. Edwards spent a few months with the Regiment there until his tour of active duty ended on June 30, 1940. He returned home and was soon informed that his commission in the Regular Army had been approved. He was ordered to report to the 22nd Infantry Regiment at Fort McClellan, Alabama.

The "Lum" Edwards memorial gavel
Passed to every President of the 22nd Infantry Regiment Society

Lula and Homer Dee Edwards Sr.

Three of the Edwards children. Left to Right: Earl, Charlene Jesse Ray
Jesse Ray Edwards was killed when the car he was riding in stalled
across railroad tracks and was struck by a train on September 7, 1923.
Ray was only three years and nine months old when he died.

Left, Mr. Vowell—right, Earl as a teenager in the store in Cruger.

Earl on the right in uniform at Mississippi State University instructing R.O.T.C. students on the Model 1917 Browning heavy machine gun.

Earl Edwards on a motorcycle. Photo believed to have
been taken while Earl was with the 11th Infantry.

2. The 22nd Infantry Regiment

Edwards began his service with the 22nd Infantry Regiment on July 1, 1940, the date of his commission as a 2nd Lieutenant in the Regular Army. He was delighted to be assigned to the 22nd Infantry since it was not far from home and many of his former classmates from ROTC at Mississippi State were also assigned to the Regiment. His first assignment was as Adjutant of 1st Battalion (only formed the month before) and commanded by Lieutenant Colonel Herbert W. Schmid, a career officer who had been in the Army since 1917. Schmid and Edwards formed a friendship beyond normal commander and assistant and which was more like father and son. Officer pay was good enough to enjoy life and Edwards bought himself a brand new automobile. It was a shiny black four door Chevrolet with radio and white side wall tires. Life in the 22nd at Fort McClellan was easy and pleasant.

Edwards described it: "Life was good. We had cars, a little extra money above our living expenses, there were girls to be dated, parties to be attended, tennis to be played, swimming pools to splash in, endless bull sessions to enjoy, and on and on. Our duties were relatively easy and interesting; we were off on Wednesday and Saturday afternoons as well as five o'clock each day." [6]

In his capacity as Adjutant, Edwards served on garrison duty at McClellan including turns as Officer of the Day (O.D.) and attended classes in military law, learning the procedures of Army courts martial. One experience as Officer of the Day was recalled by Edwards:

"…in turn each of us had to serve as "Officer of the Day". This meant you were in charge of the security of the post, the post guard house, and had to deal with any security problem that came up over the 24-hour period. You had to sleep, if you got any, with your clothes on so that you could dash off to solve the problem at a moment's notice. One night when I was on duty about 10 o'clock P.M., I received a call from Col. Schmid from his quarters. He said he was hearing gunfire over in the warehouse area of the post. I jumped up, got in my car, and sped over to the area. I knew I was supposed to have some walking post guards in the area. I rode around for a bit but couldn't find one. In those days, we were not issued pistols but wore the old-fashioned saber, if you can imagine that. It was dark, lonely, and scary and I couldn't find a guard. What if he had been killed and his killer was still around, and I only had a saber to defend myself with? I was growing more and more apprehensive, so I decided to go to the guardhouse and get some help and come back. As I rode out of the area on a hillside road, a burst of gunfire erupted right outside my car. It scared the living daylights out of me. I stopped the car and looked around and saw that there was an outdoor movie going on showing an old shoot 'em up movie. In great relief, I went back to my room and reported the facts to a much relieved Col. Schmid." [7]

On another occasion, Edwards was surprised to meet someone he knew from back home in Cruger. As Officer of the Day, at the beginning of his 24 hour tour of duty, Edwards had to check the stockade to make sure all prisoners were accounted for. When calling out one prisoner's name, Edwards was surprised to find it was a young man he knew from back home, whose nickname was "Scrappy". After speaking with the man, Edwards learned that in the Army "Scrappy" was almost always in trouble of some kind. While he was AWOL, he had been picked up by local police and turned over to the Army. Edwards gave the young man some comfort items and a small amount of money. The two men never saw each other again.

Edwards tells of his early experiences serving as an Army "lawyer" and taking part in courts martial:

"Soon after an officer is inducted into the Army, he is required to attend a school of law using the Army court martial manual as the text. Then the Army ran its own system of justice since it had jurisdiction on Army posts. If the offense is committed outside the post, the soldier can be tried by civilian authority if they so desire. So officers are assigned as prosecutor, defense attorney, or required to serve on the jury as the case might be. I had more than my share of this duty as my career went along.

"The first time I was assigned as a defense attorney for an AWOL case. These are usually very simple. If you are entered in the company morning report as AWOL, then you are ipso facto AWOL. No other evidence needed.

"So I interviewed my prisoner in the guardhouse and he told me he was certainly guilty and just wanted to plead guilty and get it over with. So this is what I did. Boy, was I naive! When the case was reviewed by old Col. Albert S. Peake, the regimental commander, all hell broke loose. He gave me the dressing down of my life. I learned that you never admit guilt. Doing that has some connotation in law that I never quite understood. He explained that if you do that you are sort of implying not only that you did it, but that you did it with malice aforethought, and had no regrets. Whereas, if you plead "not guilty" you are implying that you really didn't mean to do it, and certainly wouldn't do it again or something like that. Sounded like junk to me then and does to this day. (But you will notice that unless it is a plea bargain case, all defendants in civilian courts plead "not guilty" even if they are obviously guilty as hell.)

"Anyway, I learned my lesson. My very next case was an AWOL. I interviewed the prisoner and he said that he just did it to avoid some maneuvers and to be with his girlfriend. But this time I knew the drill. I drove down to Sylacauga, Alabama where he stayed with his girlfriend. I interviewed everybody who knew him and got statements testifying to his exemplary character and conduct. Most anybody is willing to exaggerate a little to help a poor young soldier in trouble. The local preacher had him in church often and praised him as a person of excellent character. I left with a note-book full of wonderful character references. Then I contacted his mother in Illinois. It turned out she was very poor, had a house full of small children, quite often unable to get even enough milk for her babies. She was without a husband and this soldier, her oldest child, was her only means of support. He said he often sent her money. She wrote me a long letter with tear spots on it attesting to all this. It really was a heart-breaking story, but perhaps a mite exaggerated.

"So when his trial came up, I was ready. I did not let him take the stand based on his constitutional right to remain silent (he would have ruined me!) and I saw to it that he sat quietly and looked very sad all the time.

"In the closing statement, I gave quite a performance. I pictured a poor young boy away from home for the first time, deeply in love and somewhat confused about his duties and responsibilities. Also I graphically depicted the conditions in his home as ably described by his mother. The jury bought it hook, line, and sinker. There were tears in some eyes. He got the required mandatory sentence of six months, but it was suspended. They did not even require him to forfeit his pay while he was AWOL, which was always done. This was so he could keep sending money home to feed his younger brothers and sisters. He was simply returned to duty, in effect with no punishment at all. This was unheard of, and I was a little concerned that this wasn't going to go down well in review.

"Well, all hell hit the fan again. This time it was the jury that was called to the Colonel's office. I heard he chewed them out for about half an hour, read them pertinent parts of the court martial manual, and a few other documents. Soon, I heard by the grapevine he was going to call me in again for my part in this miscarriage of justice. I wondered if I would ever get this procedure right. Anyway, he didn't call me, and that episode gradually blew over. The only result was that every AWOL soldier in the guardhouse requested me as his defense attorney. Fortunately for me, I was transferred to another post shortly thereafter." [8]

In the fall of 1940, Edwards received orders to attend a communication school at Fort Benning, Georgia. The purpose of the course was to learn how to set up telephone and radio communication networks in a combat zone. His roommate at the school was Lt. John Dowdy, who would go on to command 1st Battalion and was killed in action in Germany in 1944. Dowdy and Edwards became close friends and in his memoirs Edwards called Dowdy "…probably the best and most steadfast friend I ever had." Just before D-Day, when Edwards was given command of 2nd Battalion, he asked for and got Dowdy as his Executive Officer. Edwards would attend the funeral when Dowdy's body was returned to the United States from Belgium in 1949. Before the communication course was over, the 22nd Infantry Regiment moved to Fort Benning with the 4th Infantry Division. The Division now included the 8th and 12th Infantry Regiments along with the 22nd Infantry Regiment. The 22nd was quartered in brand new wooden barracks in what was called the Harmony Church area of Fort Benning. At the end of the course, Edwards resumed his position as adjutant of 1st Battalion.

Soon after, Edwards was re-assigned to Headquarters Company as the Regimental Communications Officer. He was authorized a jeep in his duties and came up with a novel way to lay wire for the field telephone system which was used as the primary means of communications on the battlefield. The wire had previously been laid by men on foot, stringing the wire from hand carried spools. Edwards devised a way of mounting the spools on his jeep and playing out the wire behind his jeep, laying the wire by vehicle instead of by foot. This was in keeping with the 4th Division's experiments in becoming a "motorized" Division. He was immediately reprimanded by Colonel Albert Peake (commander of the 22nd) for altering a piece of ordnance equipment without proper authorization.

However, Edwards' work with the jeep and the wire had come to the attention of the 4th Division Commander who asked Edwards to demonstrate his idea to the Division Staff. After the demonstration, the 4th Division Commanding General ordered 12 more jeeps to be outfitted the same way for the next field exercise and the wire laying jeep became an integral part of the Division.

In the late summer of 1941, Edwards was given his first command, when he was assigned as Commander of Company B, 1st Battalion 22nd Infantry. The Regiment had a large number of older Reserve officers with the rank of Captain in positions as Company Commanders. The decision was made to replace them with younger officers who were more in tune with the rapid modernization of the Army.

In the summer of 1941, the 22nd Infantry took part in the large Louisiana maneuvers. Moving by truck and halftracks from Georgia to Louisiana, Edwards learned the complicated art of motorized movement. He molded Company B into a highly trained and led unit, out-performing the rest of the Companies in the Regiment and capturing the attention of the Regimental and Divisional Staffs, earning commendations for his Company's performance. In his memoirs, Edwards related two incidents during that time which he considered to be excellent learning experiences for him:

"One morning before first light, I got a call from Regiment asking me how long it would take me to have my company at a crossroad about two to three miles away. I got the impression they wanted it done as fast as possible. Under normal procedures, this would take at least 30-45 minutes. You had to wake everybody up, get them dressed, police the campsite, and get loaded up in your vehicles before you moved out. But, on a sudden impulse, I said 15 minutes. They said for me to do it. So I raced around in the most urgent manner and ordered each platoon sergeant (I didn't have any officers in the company at that time) to simply grab their clothes, shoes, rifles, and so forth and load up and finish dressing in the truck on the way. I already had the vehicles lined up ready to move.

"The second I saw the last man in the trucks, I moved out in the lead vehicle. We made it right on the dot. At the crossroad there were several officers from Division Headquarters. They had made this a test case to see how fast we could load up and move out. The next company in line took the usual 30-45 minutes. They had simply followed the normal procedure of getting properly dressed and inspected before moving out. No one could understand how I got there so fast, and I didn't explain. It was dark, so whether the men were properly dressed was not apparent.

"The second incident took place a few days later. In those days, the Army was trying to convert from foot to motors as a way of moving about the battlefield. The transition proved most difficult. We had to develop and learn a whole new way of doing things.

"One of the problems was how to identify the unit in a column of vehicles going by without having insignia painted on the sides which would give their identity away to any spy who happened to be standing by the road. One idea was for each officer or platoon sergeant who would be riding in the cab beside the driver to have a paddle with the unit insignia painted in bold figures on it, ready for display when they spotted an officer at a checkpoint. They were to be alert and flash the paddle in a way so that the officer would know that, for example, Company B, 22nd Infantry, 4th Division, was passing by. The problem with this procedure was that often in long, hard moves the designated officer or sergeant would go to sleep and as the column passed by, no one would notice and flash the paddle to him. We were regularly chewed out and lectured on this point, but it was hard to lick. In a long, hard all-night move, it is hard for people to stay awake and alert all the time.

"On this particular occasion, the column stopped motoring about daylight. I got out of the lead truck and ran back up the column to check on the alertness of my platoon sergeants. Sure enough, they were all fast asleep. I jerked the door open and chewed them all out unmercifully, threatening to reduce rank if this ever occurred again (I reminded them of their paddle duties.) In a few minutes ,the column started to move again. Would you believe that within 100 yards there was a checkpoint with the Division Commander and staff in attendance. Now 1 know you will believe that every Company B Platoon Sergeant showed his paddle in fine fashion, and we were the only ones who did. On our return to camp, the Regimental Commander called me to extend his and the Division Commander's commendation."[8]

On October 10, 1941, Edwards was promoted to 1st Lieutenant in the Army of the United States (AUS). Promotions in the AUS were temporary yet had all the privileges and command responsibilities that went with the rank. A few months after Edwards assumed Command of Company B, Japan attacked Pearl Harbor and the United States entered the war. That Sunday, December 7, 1941, Edwards and a few buddies had gone to a movie in downtown Columbus, Georgia. When they heard the news of the attack, they returned to camp and that night Company B was ordered to station guards at a number of bridges in Southern Georgia. The fear of sabotage was an understandable reaction to the Japanese surprise attack but proved to be unfounded and within a few days all troops were back in camp.

On December 20, 1941, the 4th MotorizedDivision moved to Camp Gordon near Augusta, Georgia. First established in 1917, Camp Gordon was undergoing a huge reconstruction and enlargement, and the 4th Division was the first unit to be stationed at the new camp. The new camp commander was Colonel Herbert W. Schmid who had previously been Commander of 1st Battalion 22nd Infantry and under whom Edwards had served as adjutant when Schmid commanded the Battalion.

As 1942 began, the 22nd Infantry was still motorized and was equipped with halftracks in preparation for being sent to North Africa to take part in the desert warfare there. The North Africa assignment never came to fruition and the 22nd Infantry, along with the other units in the 4th Division, remained in the States to train new cadres who then could be sent to newly formed Divisions. On February 1, 1942, Edwards received a promotion to Captain (AUS).

In 1942, Edwards was sent to Fort Benning to take the Infantry School Basic Officers Course. Though this course is not listed in his record in the Army Registers, it lasted for three months. Upon returning to the Regiment after the course, Edwards reported to the new Regimental Commander, Colonel Hervey A. Tribolet. After examining Edwards' commendable record, Tribolet assigned him to take command of Company L, 3rd Battalion 22nd Infantry, considered to be the worst Company in the Regiment. Company L had a bad training record, poor discipline, low morale, and a high AWOL rate, and Tribolet wanted Edwards to rectify all of the Company's problems.

For Edwards it was a beneficial experience, giving him the opportunity to take charge as a forceful Commander, learning ways to make order out of chaos, instill discipline, mold a bad company into a good one, and thereby bring the morale of the unit back up to acceptable levels. It was a good training ground for Edwards and would prove invaluable in forming the kind of leadership he would exhibit with the 22nd Infantry during the later battles in Europe.

With the huge expansion of the Army in 1942, there was a shortage of officers and Edwards not only commanded Company L, he was also the only officer in the Company. The Regiment found itself without a Commander for 3rd Battalion and Edwards, as the senior officer in the Battalion, was called upon to assume Command of the Battalion. He worked both jobs at the same time, serving as Battalion Commander in the morning and Company Commander in the afternoon. He found himself in the unique position of issuing orders to himself in the morning as Battalion Commander, and then carrying out those same orders in the afternoon as Company Commander.

After a few months, the 3rd Battalion received a Major to take over command of the Battalion and Edwards was assigned by Colonel Tribolet to become the Regimental S-3 Operations Officer. Edwards had been impressed by a junior Lieutenant recently assigned to 3rd Battalion, Glenn Walker, and asked for Walker to be made his assistant. Thus began a long friendship between Edwards and Walker which lasted all their lives.

On January 30, 1943, Edwards was promoted to Major (AUS). In April 1943, the 22nd Infantry moved to Fort Dix, New Jersey. That same month, Edwards returned to Fort Benning to attend the Infantry School Officers Advanced Course. Upon completion of the course, he was assigned as Commander of 3rd Battalion 22nd Infantry at Fort Dix. On July 1, 1943, Edwards received his official promotion to 1st Lieutenant in the Regular Army. By this time the 4th Infantry Division had lost its "motorized" designation and became, once again ,simply an Infantry Division.

Since the Division and its components were now planned to be part of the assault force of the amphibious invasion landings in Western Europe, the entire outfit was moved to Camp Gordon Johnson, near Carabelle Beach, Florida, in September 1943. There the 22nd Infantry underwent amphibious landing training on a desolate beach in an underdeveloped area of Florida, south of Tallahassee. Edwards missed most of this training because from September to November of 1943 he attended and graduated from the Command and General Staff School, General Staff Class #15, at Fort Leavenworth, Kansas.

On December 1, 1943, the Regiment moved by rail to Fort Jackson, South Carolina to prepare itself for staging overseas to Europe. Edwards was relieved of his duty as Commander of 3rd

Battalion and was assigned to be the S-2 (Intelligence) Officer for the Regiment. Colonel Tribolet detailed Edwards to select two other officers and together the three of them would depart for England and become the advance party for the 22nd Infantry. On January 1, 1944, Major Edwards, Captain Neal, and Lieutenant Lawson W. Magruder left New York on the RMS Queen Elizabeth, as the first Soldiers of the 22nd Infantry to set sail for England. The Queen Elizabeth was one of the largest passenger liners of its day and had been converted as a troopship for the duration of the war.

Relying on its very high top speed and radar, the ship usually made the Atlantic crossing alone, without convoy or escort. On this voyage, the Queen Elizabeth carried nearly 20,000 troops. Four days after leaving New York, the ship landed in Scotland and Edwards and his party were moved by train to Devonshire, an area in southern England.

The British assigned a civilian to work with Edwards in finding quarters for the Regiment. The three Battalions of the 22nd would be located at one village and two abandoned Army camps. The 1st Battalion, under LTC John Ruggles, would be billeted in the town of Newton Abbot. The 2nd Battalion, under LTC John Williams, would be located at Denbury Camp. The 3rd Battalion, under LTC Arthur Teague, would be stationed at South Brent Camp. The Regiment landed in England on January 29, 1944 and the next day the individual Battalions moved to their assigned locations.

For the next four months, the 22nd Infantry was engaged in training and preparation for the D-Day invasion. Practice landings were made on English beaches, usually by loading the troops onto landing craft at the beach, taking the craft back out to sea, and then coming in for a landing at some different part of the beach. On one such occasion, in April, 1944, Lieutenant General Omar Bradley was observing the landings made by 2nd Battalion. Bradley was in command of 1st Army, of which the 4th Infantry Division and 22nd Infantry Regiment were a part. The performance of the Battalion that day must have been quite bad, as Bradley relieved the 2nd Battalion Commander on the spot, labeling him as incompetent. After a meeting with Division leaders, Colonel Tribolet informed Edwards that he would be the new Commander of 2nd Battalion.

Edwards was faced with shaping up another poor performing unit, however with the invasion less than two months away, he had very little time to do so.

Staff Sergeant Charles A. Mastro was the operations Sergeant for 2nd Battalion when Edwards assumed command. He recalled the Battalion Headquarters NCO's first meeting with their new commander:

After the Slapton Sands dry run LTC Williams was relieved, and the Battalion was taken over by Major Earl (Lum) Edwards.

His first meeting with the Bn Hq NCO's brought out his compassion for the troops. He said to us among other perfunctory messages, "Sometimes you will call me an SOB, and then at times I will do the same to you, and aside from that, I think we can get the dirty job done." [9]

One of the first official duties Edwards had to perform as a brand new Battalion Commander, was to present his Battalion for inspection during a visit to the 4th Infantry Division by General Sir Bernard Montgomery in April 1944. Montgomery was in command of all ground forces destined for the invasion of Normandy and had come to Devon where the 22nd Infantry had been assembled for his inspection.

Don Warner who joined the 22nd Infantry after D-Day during

the drive across Europe and became good friends with Earl Edwards recalled an exchange between Montgomery and Edwards:

General Sir Bernard Montgomery, while inspecting troops of the 22 Infantry in England prior to D-Day, asked various staff officers where they were from in the "Colonies." The officer being asked would reply by giving the name of his residence state. For example: When Lieutenant Colonel John F. Ruggles was asked this question by General Sir Bernard Montgomery, his reply was, "the North East Kingdom. Vermont, Sir!"

Major Earl W. Edwards replied by answering, "From the Heart of Dixie, the land of black-eyed peas and grits. Mississippi, Sir!" General Sir Bernard Montgomery was caught off guard, momentarily shocked and stunned by Major Edwards' reply. A brief second was required by the general in order for the bits and pieces to settle. Upon regaining his composure, General Sir Bernard Montgomery replied, "Gad, I thought that was a river!" [10]

On May 14, 1944 Edwards sent his mother a special Mother's Day greetings, prepared by the Regimental Chaplain, Captain Bill Boice. Boice had typed out a form letter that the troops could send to their mothers and had copies made and distributed throughout the Regiment. Each page had an invitation to a special Sunday service at the top and a loving letter to Mother at the bottom. All a Soldier had to do was tear off the bottom half of the page, which was comprised of the letter to Mother, sign it and mail it home. Boice had thoughtfully composed a beautiful glowing tribute to Mother and had taken the load of writing away from the Soldiers whose thoughts were occupied with the coming invasion. Edwards signed his copy "Your son Earl." Here is the letter:

Mr. 22nd Infantry

England May 14, 1944

Dear One,

Today is Mother's Day and I am thinking of you. Somehow it is hard to put into words how I feel about you, but there is a language spoken which our hearts have always understood.

You are very dear to me. You have taught me the things that are right and good and your love and your confidence in me have given me the courage I needed to live as a man. I can't put into words how much I love you for the numberless things you have done for me, for the countless blessings you have brought into my life, but even across the thousands of miles that separate us, I can look into your eyes and smile, and you will know, deep in your heart, what I mean.

God bless you and keep you and remember always that I am thinking of you and that time and distance will only serve to make me miss you and love you more. And when I dream tonight, it will be of the gentleness of your being, of the faithfulness of your prayers, of the loveliness that is you.

Always, [11]

As Battalion Commander, Edwards would hold command over Companies E, F, G and H. In addition to getting 2nd Battalion into a state of readiness for the upcoming assault, Edwards spent a great deal of his time studying a mock-up of the beach on which his Battalion was to land. He was provided with a Nissen hut in which the mock-up was installed, and he was the only one allowed to enter that hut. Daily aerial photos taken of the beach and inland area were brought to him and he studied them all, planning what his moves would be once the invasion got underway, in order to achieve his Battalion's objectives. As the 6th of June approached, Major Earl Edwards was as ready as he could be to take 2nd Battalion, 22nd Infantry ashore into Festung Europa.

Before he could assault into France, Edwards had another problem he had to deal with. A few days before the actual landing he suffered from hemorrhoids and was in a great deal of pain from them. He found his ailment gave him considerable trouble in walking. He asked his roommate, Tom Keenan, to report his condition to Colonel Tribolet. He was sure his medical situation would prevent him from making the landings. In his memoirs, Edwards described the resulting events:

"Later in the day an ambulance drove up to our quarters and two medics came in and said that I was to come with them. They helped me to the ambulance where I found Col. Tribolet and the Regimental Surgeon, Dr. Kirtley. We drove away and soon parked in front of a U.S. Field Hospital. Col. Tribolet and Dr. Kirtley went in and soon returned. We then drove awhile and parked in front of another hospital.

"The same thing happened. So we drove on to another. This time some medics came out and carried me in and within a short time I was operated on. After an hour or so of recovery, I was carried back to the ambulance which returned me to my quarters, and I was placed again on my bunk. All this time, no explanations whatever. I later learned that Col. Tribolet was determined that I would command the 2nd Battalion in the invasion so he and Doc Kirtley decided to take me to a hospital and ask if they would operate on me and immediately release me to their care. If the answer was "no" they simply carried me to another hospital and so on until one agreed.

"So I went into the landings with a large wad of cotton taped to my rear. Fortunately, the salt water and a few artillery rounds cured me. I don't remember ever thinking about it after we landed." [12]

At approximately 0630 on June 6, 1944, the first wave of the 4th Infantry Division landed on Utah Beach, Normandy, France. Two control vessels were disabled, and drifting currents and the smoke and confusion of the aerial and naval bombardment caused the Division to land at the wrong spot, a few thousand yards away from the planned landing area. At 0745, the first element of the 22nd Infantry, which was 3rd Battalion, came ashore. At about 1000 hours, 2nd Battalion, 22nd Infantry led by Edwards landed on the beach after crossing the Channel in LCI's (Landing Craft Infantry). Edwards' command consisted of Headquarters and Headquarters Company 2nd Battalion, Companies E, F, G, and H, and were spread out among four different LCI's. For weeks prior to the landing, Edwards had studied the beach mock-up and updated daily reconnaissance photos of the area and had memorized every inch of the terrain features.

As he approached the beach in his landing craft however, Edwards realized he could not recognize a single feature of the landscape. Upon exiting the landing craft, Edwards saw Colonel James Van Fleet, the Commander of the 8th Infantry Regiment and quickly went over to him. He asked Van Fleet if this was the right beach and was told it was. Edwards knew his mission, as Commander of 2nd Battalion, was to drive straight inland, join with any paratroopers he might find, and form a defensive perimeter to guard the beach from any German counterattack.

He led his Battalion eastward, away from the beach, and had gone about a mile when a runner from Van Fleet caught up with him. The runner informed him that the assault had landed on the wrong beach and that the route he was taking inland was the wrong one.

Edwards then had to deal with the first major decision he would make as Battalion Commander on the soil of France:

"I was faced at this moment with what I've always said was one of the most difficult decisions of my young life. Should I try to turn my battalion around and go back to the beach and try to find my correct route (this would violate every rule I had been taught) or should I just continue straight ahead and risk getting caught up in a fight and fail to get to my proper objective? After a few minutes thought, I decided on the latter as being the lesser of two bad solutions. Luckily, all turned out well. We were able to move inland, slide to the right, and get to our proper objective in time. Lady Luck smiled on me that day." [13]

As he started out, that objective was still many miles to the northwest and, due to being landed at the wrong spot, with strong enemy control of the roads inland, 2nd Battalion was forced to move through an area that had been flooded by the Germans. It took 2nd Battalion about seven hours to wade through about two miles of swamp and marsh before they reached dry ground near Ste. Martin-de-Varreville. They pushed on a bit further to Ste. Germain-de Varreville, relieved the 502nd Parachute Infantry, and set up defensive positions for the night.

The 22nd Infantry had been assigned the difficult mission of silencing two formidable German artillery batteries, one at Azeville and the other at Crisbecq. At 0700 on the morning of June 7, the 22nd Infantry began the assault on these two strongly defended batteries. Edwards led his 2nd Battalion against the battery at Azeville while 1st Battalion attacked the battery at Crisbecq. Both batteries were made up of massive concrete blockhouses holding the guns, with underground storage facilities connected by trenches and ringed with barbed wire and minefields. The immediate approach to the Azeville battery was protected by a series of concrete pillboxes manned by snipers and machine gun crews.

Edwards and his Battalion inched their way under heavy fire toward the Azeville battery for several hours until a German counterattack forced them to pull back to their original starting point. Cannon Company of the 22nd Infantry fired its 105mm howitzers to try to break up the enemy attack. In a period of about ninety minutes, Cannon Company expended 800 rounds and halted the Germans. By 1650 hours in the afternoon, 2nd Battalion was in bad shape. Edwards requested a resupply of ammunition for their M1 rifles and rifle grenades. Artillery fire was ceased while the Navy began firing in support of 2nd Battalion. Shells from the ships offshore were landing within 50 yards of 2nd Battalion's forward positions. Around 1900 hours, Edwards directed Company G to move up and resume the attack toward Azeville. Resistance was strong and 2nd Battalion was prevented from taking Azeville that day.

While 2nd Battalion was in the area near Ste. Martin-de-Varreville, Edwards and his Battalion Operations Officer, Captain Tom Neely, were walking down a trail checking on the positions of the Battalion. It was not quite dark yet, but dusk had made visibility difficult. A short burst of automatic weapons fire hit Neely as he walked beside Edwards. The fire came from a paratrooper who was stepping across his machine gun position and apparently had kicked his gun, setting it off. Neely could not be evacuated because the Germans still held parts of the road that would have been used for an ambulance. There would have been no place to evacuate him to anyway, as the hospitals and surgical teams had not yet come ashore. Neely died several hours later.

Staff Sergeant Charles Mastro was Neely's right hand man as the Operations Sergeant for 2nd Battalion. He recalled Neely's death:

"Later that evening, Maj. Edwards and Capt. Neely wanted to go down a lane to observe and make sure we were in the right area and moving in the right direction. Capt. Neely told me to give him the map case and said, "Maestro" stay back—three is a crowd." A little later, information came back that Capt. Neely had been wounded. Our Medical Officer, Capt. Humm, I believe, did all he possibly could under the present situation, but Capt. Neely died in the early a.m. A tragic incident—shot by friendly forces, apparently by a scared paratrooper. The map case had evidence of where the rounds had entered near the handle. I felt that my life was spared by not being next to Capt. Neely. He was a fine officer, and I thought a lot about him." [14]

By 2230 hours, 2nd Battalion had ceased its attack and dug in for the night. Edwards reported his losses to headquarters as four killed, 38 wounded and four missing. On June 7, at the end of the day, 2nd Battalion's five Companies measured a total strength of 854 men. The push to take the battery at Azeville would resume the next morning. Edwards ended the night by telling headquarters that an aerial bombardment of the German positions which had been cancelled that night would most surely be needed before they started out in the morning. During the night, 2nd Battalion received mortar and artillery fire from the Germans.

On June 8 between 0800 and 1000 hours, Edwards had 2nd Battalion moving forward again toward Azeville against stiff German opposition. Around that time, he lost contact with Company F which was having a difficult time moving forward. Company E had captured 35 prisoners but was receiving heavy fire from German mortars. Edwards called headquarters for help against the mortars and was told that a platoon of 4.2 inch mortars was being sent to him to provide a counter-fire against the Germans. Edwards' command post was taking sniper fire at that time and headquarters could hear the sniper fire over the phone as Edwards asked for support.

From 1100 to 1200 hours, gunfire from US naval ships were hitting the Germans about one mile beyond 2nd Battalion's positions. Companies E, F, and G were under attack and not doing well. From 1200 to 1300 hours, Edwards with 2nd Battalion, Dowdy with 1st Battalion, and Colonel Ruggles at Regimental Headquarters had a long phone discussion as to where artillery support could best be used. The German resistance in the attack against the batteries at Azeville and Crisbecq was strong, and much discussion took place about how best to press the attack. Edwards recommended to headquarters that he could bypass the Azeville battery, since the attack did not seem to be working. He related the fact that he had lost support from the tank destroyers of the 70th Tank Battalion who had apparently received orders to pull back. He maintained that flame throwers and demolitions, as well as tanks, were needed to reduce the Azeville battery and was told that tanks working with the 3rd Battalion were being sent to him.

The Germans were also employing Nebelwerfer rockets against 2nd Battalion and the destruction and the psychological terror from the rockets was having a devastating effect on his men. Edwards also had to request that supporting artillery fire be lifted from Company G's area as the rounds were falling short and dangerously close to Company G's positions.

By 1630 hours, headquarters realized that the attack against the German batteries had stalled for a second day and notified 2nd Battalion that it was to take up the same positions as the previous night. By 1700 hours, the Battalion was moving into those positions. From 1900 to 2100 hours, Edwards reported to headquarters that his battalion was receiving very little fire from the Germans. However, between 2100 and 2200 hours, 2nd Battalion again started to take rocket fire. American artillery began a counter fire, but Edwards had to request it be lifted, as shells were falling right in Company G's lines.

Around 2200 hours, Edwards was given the contact patrol line he was to establish, and he remarked to headquarters that his line was stretched pretty thin. He was notified that the attack would begin again the next morning with aerial bombardment, artillery, and naval gunfire and a Company of tanks from the 70th Tank Battalion in support.

The next morning, June 9, a decision was made by 4th Infantry Division headquarters to abandon the attack on Crisbecq. Edwards' 2nd Battalion was relieved of the attack on Azeville, that attack mission now being given to 3rd Battalion. After an intensive artillery barrage on the Azeville position, 3rd Battalion moved against the fortifications with tank support. By 1400 hours, Azeville had fallen. All three Battalions of the 22nd Infantry were then formed into Task Force Barber, the Task Force being named for and commanded by the 4th Infantry Division Assistant Commander, Brigadier General Henry A. Barber. Also part of Task Force Barber was one Company of the 899th Tank Destroyer Battalion and elements of the 746th Tank Battalion. The Task Force was directed to move between Azeville and Crisbecq and take the town of Ozeville, to the north of the German batteries, at the eastern end of the Quineville ridge.

At approximately 1630 hours, Task Force Barber began the assault on Ozeville. Halfway to its objective, the Task Force was met by heavy German resistance near Chateau de Fontenay. With darkness rapidly approaching, the Task Force was obliged to stop and dig in for the night.

The morning of June 10 found 3rd Battalion continuing the attack toward Ozeville with tank support and 1st Battalion attacking the town of Fontenay-sur-Mer, about 1,000 yards to 3rd Battalion's right. Edwards was directed to send his 2nd Battalion against the enemy positions near Chateau de Fontenay, some 2,000 yards to 3rd Battalion's rear. By 0630, Edwards had his Battalion on the attack but by 0800 the movement stalled as the Battalion became pinned down by heavy machine gun grazing fire. By 1000 hours, German rockets began hitting in 2nd Battalion's area and Edwards requested Division Artillery to fire on the rocket positions. Headquarters of the 22nd Infantry cancelled the artillery fire as it deemed that 2nd Battalion was too close to the rockets' position, and ordered Edwards to withdraw his Battalion so that the enemy emplacements at Fontenay-sur-Mer could be hit by aerial bombardment.

At about 1200 hours, Edwards began pulling his Battalion back to its jumping off point but heavy German fire caused 2nd Battalion's withdrawal to become disorganized. About seventy men of Company F became stranded and left behind. One of those seventy men turned up later as a prisoner of the Germans when Cherbourg fell. General Barber directed Edwards to reorganize his Battalion and take stock of its condition. At about the same time, Colonel Ruggles at 22nd Infantry Headquarters informed Edwards to be prepared to defend the 22nd's right flank. For the next few hours, 2nd Battalion attempted to reorganize and put itself back together.

The following statements describe the above action in which the men of Company F were lost during 2nd Battalion's attack on Chateau de Fontenay. The statements come from an interview conducted with Captain Thomas C. Harrison and Captain Floyd F. Daniel at 2nd Battalion headquarters on June 14:

"Company F was on the left, Company G was on the right, and Company E was behind Company G. At that point, a lot of small arms fire was coming from Company F's left flank. Company G also received automatic fire. Company G reached the park southwest of the Chateau and Company F reached the east side of the Chateau buildings. Capt Harold D. Fulton, Commanding Company F was killed at this point. As Company F got to the field east of the buildings, the Germans opened up with violent ambush fire. Two platoons were pinned down in the field under cross grazing machine gun fire. Company E moved up between the other two companies against the southwest side of the building.

"At this point, word was received that an air bombardment was scheduled to begin within the next hour. That was about 1200 hours. All the companies were ordered to pull back, but 70 men of Company F were trapped. Company E helped in the withdrawal of the rest of Company F. Each time the trapped men began to move, machine gun fire would open up and pin them down. We pulled back to the jump-off position which was the bombing safety limit. We then discovered that 70 men of Company F were trapped beyond the Chateau. None of these men were ever reported except one aid man who was captured by the Germans and released at Cherbourg.

"The graves of some of these men have been found at St. Mere-Eglise, including Capt Fulton, Company F commander. The original attack jumped off at 0630 hours. As word of the bombardment and withdrawal order spread, the withdrawal became disorganized. The Commander of Company E rallied his men and reformed a line to cover the withdrawal of the rest of the battalion but was unable to get to the men on the other side of the Chateau."
15

The aerial bombardment of Chateau de Fontenay never took

place and the devastation visited upon 2nd Battalion by the enemy caused that Battalion to become temporarily ineffectual. A Battalion of the 359th Infantry was attached to Barber's Task Force to continue the attack against Chateau de Fontenay. Edwards and his Battalion were told to pull back even further, to allow for an aerial bombing of the village. General Barber wanted Edwards to move out immediately, around the left flank of the 22nd Infantry, and get into position at the town of Ste. Floxel, about 2,000 yards to the east of Ozeville, to prepare to attack Ozeville the next day.

Edwards requested time to get his Battalion together and was allowed to remain in place overnight and make the movement to Ste. Floxel early the next morning.

On June 11, 1944, Colonel Robert Foster assumed command of the 22nd Infantry Regiment. Colonel Hervey Tribolet had commanded the Regiment since February of 1942 and had taken the 22nd Infantry through all of its specialized training in preparation for the invasion. He led the Regiment ashore on D-Day. By June 10 however, the 22nd Infantry had failed to advance to the locations that had been calculated it could achieve in the pre-planning for the invasion and subsequent attacks. This failure was due to stiff German resistance (and possibly un-realistic expectations on the part of the planners) and not to any incompetence on the part of Tribolet. Nevertheless, Tribolet was relieved of command and replaced by Foster.

Between 0600 and 0730 on June 11, Edwards was notified that he would get a platoon of medium tanks from the 70th Tank Battalion to assist him in taking Ste. Floxel. By noon, 4th Division ordered the 22nd Infantry to halt their advance and remain in place. Edwards and 2nd Battalion had taken their objective and were advancing west, and a liaison officer had to be sent to stop them from attacking Ozeville.

By 1800 hours, 2nd Battalion had pulled back and was in a defensive position outside of Ste. Floxel. About this time, Edwards informed headquarters that a jeep with the Anti-Tank Company commander had strayed into German lines while looking for 2nd Battalion. The officer was lost and probably killed while his enlisted driver made it back to safety. Lost along with the officer and the jeep were some documents showing the U.S. positions and a radio tuned to American radio frequencies. A couple of hours later, Edwards reported to headquarters that his Battalion had suffered a lot of casualties during the day and he requested additional medics to take care of them. While talking, Edwards had to take cover in his foxhole from intense enemy fire. He was told that at some point when the German artillery fire had lessened his casualties would be picked up. June 11 ended with 2nd Battalion dug in and sending out patrols to monitor any enemy movements.

On June 12, 1944 Company E received 12 enlisted replacements.

At about 0430 on the morning of June 12, Edwards reported to Colonel Foster at Regimental Headquarters to receive his orders for the day. The attack on Ozeville was to be resumed with 3rd Battalion making the direct assault, and 1st Battalion on its right flank providing support with tanks and artillery. Edwards and 2nd Battalion were to be on 3rd Battalion's left flank supporting the attack with mortars and anti-tank guns. The attack was to begin at noon and 2nd Battalion was to lay heavy fire on Ozeville from 1115 until 1200 hours, preparatory to the assault.

Around 1000 hours, Edwards made a reconnaissance of the ground he was to occupy in the assault, and contacted Lieutenant Colonel Ruggles, Colonel Foster, and General Barber, and convinced them to allow him to change his objective. The planned area for 2nd Battalion's positions was a natural draw and Edwards was allowed to select better ground from which to cover the attack. Once again, while communicating with headquarters, Edwards had to take cover in his foxhole from enemy fire during the conversation.

Around 1100 hours, as Edwards led his Battalion forward, he requested permission to move through the 3rd Battalion's zone with his anti-tank guns. Permission was granted and he was told he would have a platoon from the 12th Infantry to cover his own left flank. From noon to 1300 hours, 2nd Battalion's orders were changed, and it joined with 3rd Battalion in the advance on Ozeville. As yet, 2nd Battalion had encountered no opposition from the Germans. Around 1300 hours, 2nd Battalion's Company E had reached an anti-tank ditch and began receiving incoming fire. Edwards asked headquarters to notify the 2nd Battalion's supply units of the new location of 2nd Battalion's CP (Command Post) in order to get supplies up to him. He informed Lieutenant Colonel Ruggles that 2nd Battalion was running into trouble and requested permission to halt his advance. Edwards was told to keep moving.

The following passage from the 22nd Infantry's Daily Action Journal for June 12 illustrates the action at that time:

"From 1400 to 1500 hours, the 1st and 2nd Battalions are fighting like hell. Col Teague reported that he is throwing the whole dam navy on Ozeville. Maj Edwards in 2nd Battalion reported a very large caliber gun firing from their front. It is firing on the old 2nd Battalion CP. Edwards wants something to get this. He says if that gun ever zeroed in on him, there would be hell to pay. Maj Edwards told Col Ruggles lots of artillery fire is coming from Quineville. The navy is now giving it hell. Col Ruggles is trying to find out how the 3rd Battalion is getting along, but cannot get through because his phone line is out." [16]

About 1600 hours, Edwards told Colonel Ruggles that he could see elements of 3rd Battalion inside Ozeville and that 2nd Battalion was receiving artillery fire from the northwest, from the direction of the Quineville ridge. Edwards and Ruggles both tried to contact Colonel Teague, commander of 3rd Battalion, to inform him that Edwards and 2nd Battalion were ready to aid Teague by coming into Ozeville on his left flank, but both had difficulty in contacting Teague.

For the next hour, Edwards held his position and continued to be shelled by the Germans. When American artillery fire began impacting among 2nd Battalion's Companies, Edwards requested that fire be lifted and was told to keep pushing west. By 1800 hours, 3rd Battalion had taken Ozeville, and 2nd Battalion was now abreast of 3rd Battalion's line of advance. Enemy fire kept pouring into 2nd Battalion for the rest of the night and just before midnight, Edwards reported to headquarters that he was surrounded. The 22nd Infantry Daily Action Journal illustrated his predicament:

"At 2330 hours, Maj Edwards in 2nd Battalion told Maj Kenan that he believes they are completely surrounded. Machine guns on the hill are shooting down into their position and snipers are harassing the hell out of them. Edwards says he took an awful shellacking. He says that he could probably keep his position for the night but could not sustain an attack. Kenan told him the plan was for him to move under cover of darkness. Edwards says his men do not have rations yet. He says he wants plenty of artillery support. Machine gun fire is coming from his sides. He believes there are enemy tanks around. He will not leave until everything is set up for the movement." [17]

June 12 ended with Major Earl Edwards and 2nd Battalion in one bad situation.

On June 13, 1944, Headquarters Company 2nd Battalion received 16 enlisted replacements. Company F received 54 enlisted replacements. Company G received 63 enlisted replacements.

Though 2nd Battalion sustained constant fire from the Germans throughout the night of June 12, the enemy did not attack, and Edwards was able to prepare his Battalion for the next day's mission. That mission was the expected capture of Quineville, a coastal town overlooking the beach where the Allies hoped to bring in supplies necessary for the assault on Cherbourg. As the 39th Infantry made its way along the coast northward toward Quineville, the 22nd Infantry was to secure the ridge to the east of the town and join with the 39th Infantry in taking the town itself.

The plan called for Edwards to move his Battalion wide to the left, through the area of responsibility of the 12th Infantry, and secure the eastern end of the ridge. At 1000 hours on June 13, Edwards had not yet begun his attack, as he was held up by a unit of the 12th Infantry to his front who were engaged in heavy fighting with the Germans. When it was able to, 2nd Battalion moved along the Quineville-Montebourg highway toward the ridge but sustained 20 casualties in a sudden firefight. Edwards at first reported to headquarters that he was fired upon by a unit from the 12th Infantry in a case of mistaken identification, but later corrected his report when he discovered it had been the enemy who opened fire.

At 1100 hours, headquarters requested tank support from the 4th Division to aid 2nd Battalion in its movement to the ridge. When Edwards was informed there were no tanks available at the time, he replied that he was ready to go without them. Around 1500 hours, two medics badly needed by 2nd Battalion were killed by snipers as they moved up to the area. Facing stiff German resistance, 2nd Battalion's movement was slow going. By 2200 hours, Edwards was told by headquarters to establish a defensive position for the night. He would resume the attack at 0630 with tank support.

The attack for June 14 was rescheduled for 0930 with a fifteen minute barrage fired by 4th Division artillery in advance. Edwards moved his Battalion out at 0930, his objective being the two hills which made up the western end of the Quineville ridge. Though communications between headquarters and 2nd Battalion was lost several times from the early morning hours well into the late afternoon, Edwards led his Battalion steadily forward, encountering strong German opposition. By 1300 hours, Companies E and G of 2nd Battalion had taken their objectives. Edwards called for a spotter plane to locate enemy mortars which were giving Company E a difficult time and notified headquarters that large caliber guns and rockets were visible to the north of Company G's defensive lines. At 1500 hours, enemy artillery fire was getting closer to Edwards' command post, his message center was under heavy fire, and his communication with his own artillery support was out. He was suffering a high amount of casualties and requested medical support. He also asked for artillery fire upon a German battery northwest of Montebourg that was giving his Battalion grief.

Edwards and his Battalion had taken and occupied their objective, the two hills on the western end of the ridge. By 2000 hours, Edwards was directed to dig in for the night. For the first time in eight days, he was not given an objective for the next day. Major Earl Edwards and 2nd Battalion would be allowed to rest for the first time since making the landing on June 6.

On June 15, 1944, Company H received 51 enlisted replacements.

June 15 found 2nd Battalion still on the front lines but not ordered to attack. Early in the morning, some 30 Germans surrendered to elements of the Battalion. Between 1300 and 1900 hours, Edwards consolidated his lines with those of the 12th Infantry on his left flank. He was instructed by headquarters to keep in close contact with the 12th Infantry. Other than some intermittent artillery fire from the enemy, 2nd Battalion's sector was quiet.

It remained quiet throughout the day of June 16 as VII Corps formulated its plan to take the city of Cherbourg.

On June 17, 1944, Company E received 52 enlisted replacements. Company F received 16 enlisted replacements. Company G received 17 enlisted replacements.

June 17 saw Edwards tasked with searching for and destroying any German communications wires and cables that may have been left behind during the previous advances. He was notified that 3rd Battalion would be given the mission of taking Montebourg to the northeast. Edwards was instructed to place one of his Companies in mobile reserve during the attack. At approximately 1600 hours, 2nd Battalion began to be relieved of its positions by the 24th Cavalry. 2nd Battalion waited until dark to move out in the direction of Fontenay-sur-Mer where the Regiment would be assembled.

Throughout the day of June 18, all of 2nd Battalion was held in mobile reserve by headquarters. Sometime after 1600 hours, Edwards was notified that Company F from his Battalion would be detached from the 22nd Infantry and assigned to the 70th Tank Battalion for the assault on Montebourg. That Company was to have with it a section of mortars and a squad of machine guns. Company F moved out at 1800 hours to join the 70th.

June 19 saw the continuing of the push to Cherbourg. Though the enemy had four days to build up their defenses, when the two

Divisions on either side of the 4th Infantry Division began their advance, they outflanked the Germans, causing them to pull back, leaving the 4th Infantry Division relatively unopposed. When 3rd Battalion reached Montebourg, it found the city deserted. For the remainder of the day, 2nd Battalion was held in reserve. During the night, the 22nd Infantry was assembled on the northern edge of the Quineville ridge in order to bring it into position to support the 12th Infantry during the next day's advance.

At 0330 on June 20, the 22nd Infantry moved out on the flank of the 12th Infantry. The Germans disengaged and pulled back to make a stand at the Port of Cherbourg, leaving open ground in front of the 22nd Infantry. The Regiment made rapid progress that day with only sporadic resistance from the enemy. Edwards and 2nd Battalion stopped at 1800 hours and dug in for the night.

The objective for the 22nd Infantry on June 21 was to cut the highway which ran from Saint-Pierre-Église northward to Cherbourg. The attack was concentrated on an area near the town of Gonneville-Le Thiel. 1st and 3rd Battalion met heavy resistance in front of their objectives while Edwards and 2nd Battalion were held as mobile reserve for the entire day. As night fell, the situation for the other two Battalions was so dire that 2nd Battalion was brought up near them, to be in a position to rush to their aid at a moment's notice. Edwards and his men were shelled rather hard as they moved up to the front line. The night ended with 2nd Battalion still being held in reserve.

June 22 began with an ultimatum delivered to the enemy commander of Cherbourg demanding his surrender. When the deadline for replying ended without reply from the Germans, the attack on the city began in earnest. At 1300 hours, some 500-600 Allied fighter-bombers bombed and strafed the front line enemy positions. Artillery smoke which had been fired to mark the German lines drifted into the American lines and Edwards reported that the planes were shooting up his Battalion and causing casualties. He asked for and received permission to end his Battalion's reserve status and move to the aid of 1st and 3rd Battalions who were being met with fierce resistance. Enemy artillery was particularly accurate, in addition to German infiltrators who were cutting supply routes and making it difficult to bring up needed food, water, and ammunition. Tanks which should have been used in the attack were diverted to protect supply convoys coming up to the front.

From 1600 to 1800 hours, 2nd Battalion took heavy fire from both flanks, not only from small arms but also from German 88mm guns. Edwards kept pressing his men on in the attempt to reach Major John Dowdy and 1st Battalion who were taking a severe beating from the enemy. At 2100 hours, all units were ordered to halt their advances and consolidate so that supplies could be brought up and their casualties could be evacuated.

At 2200 hours, Edwards reported to headquarters that his Battalion's rear command post had been knocked out by an artillery barrage and his Company E was in the process of being surrounded. Edwards lost communication with headquarters shortly thereafter and new communications wire could not be brought in because of heavy enemy fire. Near midnight, Company E was cut off from the rest of the Battalion by the Germans and their medical section was hit hard. A runner came into headquarters and related that the 2nd Battalion command post was slowly being wiped out. Lieutenant Colonel Ruggles from headquarters took two tanks to try and reach Company E and 2nd Battalion. Some men from the 24th Cavalry went sent to Company E's aid. "Swede" Henley's Cannon Company of the 22nd Infantry even sent a patrol to try to reach Company E. By 0300 the next morning, the situation had been relieved and Company E was finally out of danger for the time being.

The plans for June 23 called for the 22nd Infantry to assist the 12th Infantry in an attack on Tourlaville, a town only about one mile from the city of Cherbourg. The 22nd was to clear out the fortified area around the village of Digosville, about four miles to the southeast of Cherbourg and thus protect the 12th Infantry's right flank. However, sustained German operations against the 22nd Infantry from enemy units stationed around Gonneville and the airfield at Maupertus near the coast kept the Regiment involved in efforts to keep the vital supply lines open. The 3rd Battalion of the 22nd was to lead the attack against Digosville from its position on the high ground of Hill 158, to the northeast of Digosville, but the Germans flanked and nearly surrounded Hill 158 and Edwards and 2nd Battalion had to be committed to clearing the enemy from 3rd Battalion's rear. This mission took most of the day, until about 1400 hours.

In addition, 2nd Battalion sent a combat patrol to its south aided by Company C of the 4th Engineers, to clear out and mop-up an area around the village of Hameau Cauchon. Considerable artillery and mortar fire was placed on the German lines in the Gonneville and Maupertus areas to cover this patrol.

Edwards was not in communication with his Company E which had sustained such a hard time the previous evening. He called several times to Headquarters for information on the Company and asked when it might join the rest of the Battalion. He was told the Company was now in good shape and would remain where it was in order to protect the Division's right flank.

With the pressure on 3rd Battalion relieved, and the successful conclusion of the mop-up operation, 2nd Battalion was now ordered by headquarters to attack northward from Hameau Cauchon and westward toward Digosville. Edwards moved his men out at 1930 hours and for more than an hour his Battalion came under heavy fire from the Germans positioned southeast of Digosville. He never reached his planned line of departure for the attack. By 2130 hours, he stopped his advance and dug in the 2nd Battalion for the night. Edwards was notified that as of 2400 hours, his Battalion would come under command of the 12th Infantry for the next day's activities. Company E was instructed that at 0700 the next morning it would be relieved of its assignment by a mobile reserve Company of the 12th Infantry, and it would then be free to rejoin 2nd Battalion.

On June 24, the three Division attack on the city of Cherbourg closed a semi-circle around the city with the 4th Infantry Division covering the eastern third of that semi-circle. Leading the 4th Division's assault was the 12th Infantry with Edwards' Battalion attached. Edwards was given the task of taking Digosville and was joined by Company K of the 12th Infantry. Edwards was prepared to have his Battalion attack alongside of Company K and its tanks, but communications between his command and the 12th Infantry were not well thought out. Edwards did not communicate directly with 12th Infantry headquarters. He received his orders from 22nd Infantry headquarters, which talked to the 12th Infantry and relayed instructions.

The main form of communications by Army elements was the field telephone. Each Regiment laid their own telephone wires across the battlefield to each of their respective organizations but not to organizations of a different Regiment. Worse yet, the two Regiments used different reference numbers for the same objectives.

At 0830, Edwards moved his Battalion out to join the 12th Infantry. He was one Company short as Company E had not joined with 2nd Battalion because it was waiting for a guide to lead it. Edwards was concerned that 2nd Battalion would be without artillery or mortar support, as the 12th Infantry support would be busy firing for their own organic units. Colonel Foster at 22nd headquarters assured him that if he did need support, Foster would get it for him. Between 0900 and 1000 hours, Company E rejoined 2nd Battalion and Edwards reported to 22nd headquarters that he was moving forward without resistance from the enemy. Around 1200 hours, 12th Infantry called 22nd Infantry and asked for the location of Edwards' 2nd Battalion. Company K of the 12th Infantry had made the attack on Digosville by itself and had taken a German position just outside of the town, supported by tanks and an airstrike by P-47 fighter aircraft.

The following passages from the 22nd Infantry journals for June 24 show the confusion and concern experienced by Edwards from being attached to the 12th Infantry:

"Maj Edwards with 2nd Battalion said he was in a mess because he was attached to the 12th Infantry. In addition to that, he was getting instructions from other people. He did not know what the hell the score was. The location numbers that the 12th Infantry gave as their objectives were different from our numbers.

"From 1100 to 1400 hours, the 12th Infantry called to say that our 2nd Battalion would continue the attack towards Digosville after relieving Company K of the 12th Infantry. Col Ruggles with the 22nd Infantry so informed Maj Edwards. Edwards wants more information about the objective. Ruggles will call the 12th Infantry and find out what is what. Division informed the 22nd Infantry that Company F of the 12th Infantry would go back to their old area. Division was told about the 2nd Battalion receiving orders direct from the 12th Infantry. Division is investigating. Gen Barton, Commander of the 4th Infantry Division, visited the CP.

Gen Barton called the 12th Infantry and said they would give orders direct to the 2nd Battalion. Much confusion has resulted from the 2nd Battalion being assigned to the 12th Infantry." [18]

By 1700 hours, headquarters gave Edwards information about Digosville obtained from prisoners. Edwards relayed that the objective was too big to take all at once, that he would take it in sections. He expected to reach Digosville and begin his attack at 1800 or 1815 hours. Nearing the town, his Battalion ran into a Company of Germans and captured six of them. The prisoners told him that the rest were willing to surrender, and Edwards sent some tanks to pick them up.

The 22nd journal of June 24 gives the following details at that point:

"From 1800 to 1900 hours, Company G gave their position to Lt Henry in the 22nd Infantry. They said the mortars would not move forward and wanted Col Ruggles to do something about it. Also, they said a prisoner told them they were expecting reinforcements in Digosville but did not know whether or not they had arrived yet. The 2nd Battalion complained to Col Ruggles that they were still having troubles with their communication link to the 12th Infantry. Said they had to go way back almost to their Regimental CP to trace the trouble. He asked Col Ruggles to do something about it. He said his wire team is having to do all the trouble shooting themselves without help from the 12th Infantry. Col Ruggles then called Capt Hickey at the 12th Infantry about it. Maj Edwards reported that the 2nd Battalion is on their objective, meaning he has captured Digosville. He has quite a few prisoners and expects to take more. They will completely search out the place and then sit down for the night. He cannot hold the whole place, so he is going to stay in part of it and then patrol the rest of it." [19]

At about 1900, Edwards reported that his Battalion was consolidating its position in the southern section of Digosville and that it had captured about 60 prisoners. He told headquarters that the northern section of the town would take longer to occupy, as the Germans had ringed the area with a lot of barbed wire, mines, and booby traps. By 2100 hours, Edwards said that Digosville was almost cleared out of enemy and that 2nd Battalion had amassed about 150 prisoners. During the entire day, the 12th Infantry (with Edwards' Battalion attached to it) had captured over 800 enemy prisoners.

On June 25, 1944, Headquarters Company 2nd Battalion received two enlisted replacements. Company E received 18 enlisted replacements. Company F received 41 enlisted replacements. Company G received 35 enlisted replacements. Company H received 18 enlisted replacements.

The 12th Infantry continued to push northward and by 0200 hours on the morning of June 25th rolled unopposed into the larger town of Tourlaville, just to the north of Digosville. At 0300 hours, Edwards was given the order to march north from Digosville and penetrate to the coast.

Sometime after 1000 hours on the morning of June 25, 2nd Battalion was released from the 12th Infantry and rejoined the 22nd Infantry Regiment. Edwards continued his advance to the coast, taking his objectives on time and meeting little resistance. By 1900 hours, Company F of the 2nd Battalion had reached the coast about a mile east of Point des Grèves, some three miles east of Cherbourg. The 12th Infantry had been chosen as the element of the 4th Infantry Division that was to attack the eastern end of the city of Cherbourg itself, while the 22nd Infantry protected the right flank and cleaned up pockets of German resistance.

As the ring closed around Cherbourg, the Germans were surrendering in large numbers as their front lines were pushed back and collapsed. Edwards reported to headquarters that he suffered only two or three casualties and captured about 500 prisoners throughout the day. He was ordered to assemble his Battalion near Hill 198 and be prepared to attack at 2000 hours. It is not known if the attack was carried out, as the daily journal for the 22nd Infantry on June 25 ends at this point.

On June 26, 1944, Company E received 10 enlisted replacements. Company F received 10 enlisted replacements. Company G received 9 enlisted replacements.

On June 26, elements of all three American Divisions were entering the city of Cherbourg and its surrender was imminent. The 22nd Infantry was thus directed to focus its attention on one of the German strongpoints which had been contained and bypassed in the advance on Cherbourg. That area of the peninsula to the east of Tourlaville, forming a triangle roughly from the village of Gonneville to the south, Cap Levy (Cap Lévi) on the coast to the north, and continuing nearly to the town of Saint-Pierre-Église to the east, was still occupied by a few thousand men of various German units. Elements of the 22nd Infantry had been facing and containing the enemy on the western and southern sides of this area during the drive to Cherbourg. The Regiment was now ordered to attack and seize the airfield at Maupertus, about three miles to the southwest of Saint-Pierre-Église.

At 1100 hours on the morning of June 26, the 22nd Infantry attacked from the west with all three of its Battalions abreast in line. 1st Battalion took several German positions south of the Maupertus area, including the village of Gonneville. 3rd Battalion captured the town of Maupertus-sur-Mer to the northwest of the airfield. At 1125 hours, Edwards and 2nd Battalion took hold of the western end of the airfield itself, but his Company F became pinned down by heavy fire from German 88's. Heavy enemy resistance prevented all Battalions of the Regiment from advancing any further. Sometime after 1200 hours, Edwards was cautioned that his men were firing into the 3rd Battalion's zone of operations. Just after 1500 hours, Edwards was asked by Major Keenan at headquarters if 2nd Battalion could push any further into the airfield. Edwards replied that his Battalion was virtually stopped by heavy fire from German anti-aircraft guns. The airfield had numerous anti-aircraft guns of large caliber protecting the airfield from U.S. air attack, but the enemy had turned the gun barrels down to fire against the oncoming American infantry.

Well after 1600 hours, Edwards relayed to headquarters that the anti-aircraft guns were still holding up progress of all three Battalions. Company G of 2nd Battalion had silenced one of the guns, but the others were difficult to locate. Colonel Foster, Commander of the Regiment, joined Edwards at 2nd Battalion's location to view the battle for himself. Upon seeing the situation firsthand, Foster sent a few tanks to aid the 3rd Battalion in its assault. By 2000 hours, Edwards reported that he was receiving intermittent machine gun and artillery fire. By 2200 hours, progress had ended for the day. Edwards told headquarters that 2nd Battalion had 24 casualties for the day and he requested that he be given a naval gunfire liaison officer to direct naval fire for him the next day.

Throughout the night, Edwards reorganized his Battalion for the attack of June 27. He was informed that his Battalion would be given a Company of medium tanks and a platoon of light tanks to aid him in his advance. By 1600 hours on June 27, 2nd Battalion had been moved by headquarters to attack on the right flank of 1st Battalion. By 1845 hours, Edwards and his men were moving onto the airfield itself. German resistance throughout the entire area was crumbling and it was expected that the Germans would surrender shortly. There were still a few pockets of strong enemy resistance, but word had been received that the German officer in charge of the entire sector was willing to surrender his forces. American artillery fire was either called off or heavily curtailed for the rest of the night.

By 0700 hours on the morning of June 28, the surrender had been verified and the 22nd Infantry began planning on how to assemble and move the large number of German prisoners expected. Edwards was notified that the Regiment's kitchens would be bringing up hot food and that any prisoners he took were to be sent to the south end of the airfield. By 1500 hours, Lieutenant Colonel Ruggles had told Edwards to keep out of any fortified areas that might be set with mines or booby traps and that the men of 2nd Battalion were to rest, shave, and generally clean up and take a breather. Edwards found a nearby chateau in which he established his Command Post. By midnight, all of 2nd Battalion except for Edwards and his Command Post had been moved to the Regimental assembly area, which was the airfield itself.

The only major activity for the 22nd Infantry on June 29 was a ceremony to hand out Silver Star Medals. Each Battalion was to furnish a Company made of one platoon from each of its rifle Companies. These composite Companies would be the honor guard for the presentation. Those Companies, the award recipients, and all Officers who could attend, were gathered at the southern end of the Maupertus airfield for the ceremony. General Raymond Barton, Commanding General of the 4th Infantry Division, presented the awards.

On June 30, 1944, Company F received 11 enlisted replacements. Company G received 12 enlisted replacements.

On June 30, the 22nd Infantry moved to a new assembly area about three and a half miles to the west of Ste. Mère Église near the town of Amfreville. The Regiment stayed in this area from

July 1 through July 6, conducting training and cleaning, repairing, and consolidating its equipment for upcoming operations. The next phase of the 22nd's action in France was to take part in the fighting in the hedgerow country in the Allied attempt to break out from the beachhead. While in the Ste. Mère Église area, the Battalion was out of direct combat, but German artillery was always a danger. Captain Jim Burnside from Company E witnessed a rather humorous episode involving Edwards:

"On a later date (I believe around St. Mere Eglise,) during a lull, Major Lum Edwards, our battalion commander, had cautiously worked his way to a shallow shell hole in the middle of a small field and squatted down to do his morning's duty. A shell landed in the same field. Lum squatted down a little deeper and pulled his helmet down a little further. Wham! Another shell hit a little closer. Lum—never rising up, holding his pants around his ankles in one hand and his helmet on with the other—did the most amazing duck waddle to safety in a nearby hedgerow, accompanied by our hysterical laughter." [20]

July 6 found the 22nd Infantry loaded into trucks and moving to a new assembly area about eight miles to the south, just west of Carentan. This would be the jumping off point for its next attack, in a southerly direction along the Carentan-Perriers road. At 2205 on the night of July 6, Edwards called headquarters to say he did not have enough yellow smoke mortar rounds and yellow smoke grenades to use to mark his front lines should he use air and artillery support. He was told to go back to the supply point and gather all he needed.

The 22nd Infantry remained in reserve the entire day of July 7, alerted to be ready to move out in one or two hours' notice, but it never received the order to actually move out. During the day, the 8th and 12th Infantry Regiments established defensive positions along a line of departure for the planned attack of July 8. The next objective of the 22nd Infantry would be the capture of the town of Periers, to the southwest of Carentan. Edwards was given the mission of breaching the German lines just below Culot, then continuing on to take the town of La Maugerie, about 10 kilometers southwest of Carentan, just two kilometers southwest of Sainteny.

After an aerial strike on the German defensive lines, the 22nd Infantry was to attack in column with the 2nd Battalion leading the assault. At 0302 hours on the morning of July 8, Edwards was informed by headquarters that the aerial bombardment would be delayed. He was told that when it did start however, he could call for more aircraft strikes if he needed them. At 0930, the bombing began. At 0945, 2nd Battalion moved out. At 1008, Edwards crossed the Line of Departure. He reported taking casualties from "S" mines but no resistance. At 1100 hours, 2nd Battalion began to meet German opposition. At 1205, Edwards called headquarters to request mortar fire to support his attack. At 1210, he reported enemy shell fire passing over his command post. Automatic weapons fire stalled 2nd Battalion at this time and Captain Moon informed headquarters at 1225 hours that heavy caliber fire was passing over 2nd Battalion at five minute intervals. Edwards continued to report the locations of his front lines and his command post at frequent intervals during the action. At 1507, Captain Moon with 2nd Battalion informed headquarters that Company G was still advancing, but Company F was pinned down and had lost one tank. At 1735, Edwards reported that his Battalion was receiving small arms fire from his front and both flanks. Progress was slow and hindered by constant German fire. At 1815 hours, headquarters instructed all units to stop at 2100 hours and dig in for the night. The attack would be resumed at 0730 the next morning. The assistant Division Commander of the 4th Infantry

Division, Brigadier General Teddy Roosevelt, Jr. visited the 22nd Infantry Commander Colonel Robert Foster at Foster's command post at around 2000 hours.

After he left, at 2018 hours, Foster called Edwards for a report of 2nd Battalion's progress so far. Edwards replied that progress had been slow because of enemy artillery fire, and he noted to Foster that 1st Battalion was also hindered by artillery and moving slowly as well. Foster advised Edwards to disregard the previous instructions to dig in and to continue pushing forward, keeping Foster advised of his progress.

According to Chaplain Bill Boice ,2nd Battalion had by this time only advanced to within 500 yards of the village of Culot, still about a mile and a half away from their objective of La Maugerie.[21]

At 2151, Major Kenan at Headquarters instructed 2nd Battalion to stop the attack and button up for the night at its present point. Edwards replied that he couldn't stop at the moment because his Battalion had been under attack by the Germans since 2130. The enemy had launched a counterattack against 2nd Battalion's front and right flank with tanks and infantry. The action was described in Bill Boice's *History of the Twenty-Second Infantry* thusly:

[About 2130, three German tanks accompanied by infantry came up the road against the Second Battalion's position. The leading tank continued up the road between Companies F and G toward the rear of the battalion. At the road junction a few hundred yards in the rear, stood an American tank which had been knocked out and was blocking the road. Lt. Colonel Wellburn, CO 70th Tank Bn., had just come up to examine this tank when he heard the German tank approaching. At the same moment, Colonel Wellburn saw an American half-track towing a 57mm gun coming down the road behind him. He stopped the half-track and said, "Uncouple that gun. Here comes a Kraut tank." The driver immediately pulled into the farm yard just behind the wrecked tank and the gun in a few seconds was uncoupled just at the corner of the gate, covering the road.

As the German tank came around the bend in the road, he saw the American tank a bare hundred yards in front of him. The German stopped and opened fire. His first round set the American tank on fire and he moved forward slightly and continued firing. Several rounds went through both sides of the turret of the American tank, and one went all the way through the tank and knocked the door off the back.

Meanwhile, the 57mm. antitank gun was loaded and ready to fire. As the side of the tank came into view, the sergeant said, "Pour it in." The first shot hit the most vulnerable point and knocked the German tank out. The 57 put five more shots in the same place. When this encounter ended, the two tanks, German and American, were facing each other 50 yards apart, both riddled with holes and their crews wounded or dead.

A second German tank accompanied by infantry had swung west into the orchard in front of Company F, opening fire on the company. At the same time, the Germans put down a heavy artillery barrage on the orchard.

Another German tank had swung east into the orchard in front of Company G. Private Hicks, with a bazooka, stood at the corner of a small house near the left flank of Company G and fired at the approaching tank. He got four hits, and on the fourth the German tank blew up. The turret was blown off, and the tank tipped over.

With two of the German tanks knocked out, the third withdrew. Company F was immediately ordered to retake the field from which they had withdrawn, and Captain Tommy Harrison, Battalion S-3, moved up to tie in the lines and prevent gaps from forming. The 44th Field Artillery, in direct support, laid down such a heavy barrage on this field and the enemy positions behind it that the smoke completely obscured the scene. Company F attacked just before dark but was stopped by heavy German fire. About 0230 the next morning, Lt. Clark and six men who had been in the disputed field throughout our bombardment, succeeded in returning to our lines.] [22]

The "Private Hicks" mentioned above was actually Private Eugene Hix from Tennessee. He was one of Edwards' men, a Private First Class from Company G, 2nd Battalion 22nd Infantry. For this and subsequent actions through July 13 when he was killed in action, Hix was awarded the Distinguished Service Cross.

Donald Nolan, serving at the time of the above action as the Company communications runner for Company G, witnessed Edwards as Wellburn maneuvered the 57mm anti-tank gun into position. He described the scene as the three German tanks made their way into the lines of 2nd Battalion:

"There was a sunken road, and on each side were fields with hedgerows. I was about a hundred and fifty yards from the Battalion CP, walking out in the field, when I heard a lot of commotion on the road. I looked through the hedges and was looking into the biggest German tank I ever saw. It was headed toward our CP on the road.

I ducked and headed back to the CP. There was Major Lum Edwards, standing in the hedgerow break, watching our antitank gun and half-track retreating down the road. What he didn't realize was that the gun was pointing the wrong way and was trying to get turned around. Major Edwards was sort of moaning that his own antitank gun was running away. When the gun finally got turned around, it put at least three 57mm rounds into the tank."[23]

At 2205 hours, Edwards called headquarters and requested that a Company be sent to the aid of his Company F who had been hard hit in the enemy counterattack. Major Kenan replied that he would send an anti-tank platoon and find an infantry Company to send to Edwards. At 2209, Kenan talked with Lieutenant Colonel Teague, Commander of 3rd Battalion, and told Teague that the situation was grave, that Edwards didn't know how many enemy tanks and infantry were in the area but one large tank was on fire and another was pulling back.

Teague committed his Company L and a machine gun section to the aid of Company F. Teague ordered Captain Howard Blazzard from Company K to find Edwards and consult with him and further instructed Blazzard to put Company K on alert should they be needed to also help Company F. Kenan further ordered that 600 rounds of white phosphorous be fired by mortar support on the road from which the German tanks had advanced.

At 2310, Edwards reported to headquarters on the night's activity. He said that 2nd Battalion had knocked out two enemy tanks and a large number of enemy infantry. He estimated the German infantry to have been Company size and relayed that the tanks were big ones with 88mm guns on them. In his diary, Major "Swede" Henley noted that the knocked out German tanks were Mark V Panther tanks. [24]

July 8 ended with the German counterattack against 2nd Battalion repulsed and Company F out of trouble for the time being.

July 9, 1944 was to become a major event in the history of the 22nd Infantry Regiment. Colonel Robert Foster was relieved of command of the Regiment. Major General Barton had replaced Tribolet with Foster and now he was replacing Foster with Colonel Charles T. "Buck" Lanham. Stiff German resistance had prevented the 22nd Infantry from achieving its pre-planned objectives in the campaign and once again the Regimental Commander took the fall for something which was beyond his control. At approximately 0800 hours, Lanham assumed command of the 22nd Infantry.

Lanham was a no-nonsense guy who accepted no excuses for failure, and this was his first combat command. He would get the credit for the success the 22nd Infantry enjoyed after the initial setbacks the Regiment experienced in Normandy. He introduced himself to the Regiment in a fiery and flamboyant way:

A field phone rang. A Captain answered and heard, "I am Colonel Charles T. Lanham. I have just assumed command of this regiment, and I want you to know that if you ever yield one foot of ground without my direct order, I will court martial you."[25]

At 0032 in the early morning of July 9, Edwards sent a radio message to headquarters informing them that the telephone land line communication wire between headquarters and 2nd Battalion had been completely destroyed, and asked Headquarters for help in repairing it. Major Kenan at headquarters replied that men from headquarters would start working on repairing the line from their end and that Edwards should send someone to start the repairs from 2nd Battalion's end, with the aim of meeting up somewhere in between.

By 0545, Edwards had received his attack orders for the day from the Division Liaison officer. By 0615, the communications wire had been repaired. At 0620, headquarters received a message that 2nd Battalion had knocked out another German tank, this one of 60 tons. At 0810, Lanham spoke directly to Edwards and informed him that the previous day's attack must be carried out again. Lanham maintained that the ground given up by 1st and 2nd Battalions on July 8 had been responsible for the failure of the Regiment to achieve its objective for that day. He ordered Edwards to outflank the enemy who had moved into the positions vacated by the 22nd Infantry and to use all the artillery support that was available to regain that lost ground. At that point, the communication line between 2nd Battalion and headquarters went out again.

Edwards moved his men out. Progress was slow but 2nd Battalion kept moving forward. At 0943, Lanham was told by Division to hold up 2nd Battalion's advance until the unit on its flank, the 331st Infantry from the 83rd Division took their objective. With communication between Edwards and headquarters interrupted by the lost land line, at 0945 Lieutenant Colonel John Ruggles, the 22nd 's Executive Officer, was sent to personally take command of 2nd Battalion and hold it in place until it could be released to once again to join in the attack.

At 1230 the communication line from 2nd Battalion to headquarters went out again. By 1300 it was repaired and working again.

Edwards resumed the attack at 1355. At 1430, he lost his anti-aircraft support when the battery was detached and sent to another unit. At 1445, Edwards was informed that all units were to be ready for an artillery barrage to precede further movement. First and Second Battalions were directed to move in behind the barrage and take the village of Les Forges. The barrage commenced at 1500. Three minutes after the firing started, Captain Tommy Harrison, commanding Company E, called the 20th Field Artillery to tell them their rounds were falling short and hitting inside the front lines of 2nd Battalion. At 1510, Lanham called Ruggles to tell him to investigate the report of short fire since Ruggles himself had not "OK'd" the report.

At 1530, Edwards called a supporting unit to inform it that 2nd Battalion was moving forward. Edwards was asked what kind of tank they had knocked out in 2nd Battalion's area. The reply was a Mark VI (Tiger tank). Edwards was told that whoever used the bazooka to destroy the tank deserved a decoration. At 1535,

Captain Harrison, with E Company, reported to headquarters that his Company was receiving machine gun fire from the houses in Les Forges. Lanham replied that he would direct 1st Battalion to assist and would have artillery fire coordinated in support. The two Battalions continued to push forward toward the villages. At 1615, the 20th Field Artillery was informed by 2nd Battalion that the short fire experienced earlier was now believed to have been enemy fire, not friendly fire.

At 1643, Captain Burnside, with Company E, informed Lanham that 2nd Battalion's front lines were on the road southwest of Les Forges, with Company E on the left, Company G in the center, and Company F on the right. By 1647, Les Forges had been taken by the 22nd Infantry with 2nd Battalion in place covering the road. The 331st Infantry had to be called to halt its firing, as small arms fire from it was being received in the 22nd's area. At 2100, the 22nd Infantry held up its advance and dug in for the night. At 2220, Lanham called for a meeting of all battalion commanders to be held at 0630 the next morning.

In a wartime interview conducted with Captain Tommy Harrison and Captain Floyd Daniel of 2nd Battalion, the two officers had this to say about the action of July 9 in the attack toward the village of Les Forges:

"Lt Clark came back about 0230 hours on July 9 with information about a platoon of Germans digging in the field that was abandoned due to heavy fire by Company F. German artillery fire on us kept up all night. The 1st Battalion had gotten one company across the creek south of Culot almost to Neuville. The remainder of the 1st Battalion pulled back to our side of the highway and also went a few hundred yards to our right rear. The 331st Infantry was to our left rear. Company E would move through Company F and take the orchards to the west of the buildings. Company E and Company G were astride the road. Les Forges was given as an intermediate objective to Company E. They jumped off on their attack at 0800 hours on July 9. From there on in, they started to advance hedgerow by hedgerow. Company E took the buildings. The 1st Battalion was abreast across the highway with the Germans falling back from hedgerow to hedgerow as we advanced. The 1st Battalion, on the right, had a hell of a time crossing the stream but made it that night. Capt French of Company E was killed on July 9. Company F turned southeast and took part of Sainteny in conjunction with the 331st Infantry. Company E had taken Les Forges which was pretty well knocked out by artillery. The 81's (American mortars) fired over 2500 rounds during July 9. We had eight medium tanks, also Cannon Company, 81 mm mortars and 4.2's. Priority of fire by Cannon Company and the 4.2's went to the 1st Battalion. About eight medium tanks were attached to 2nd Battalion. On July 9, Private Hix got a second tank with a bazooka. Hix's first two tanks were German Mark Vs. The Germans sent harassing artillery and mortar fire on us all day of the 9th." [26]

On July 10 at 0330, 2nd Battalion radioed headquarters that the unit on its flank reported five plane loads of enemy paratroopers had landed behind 1st Battalion. For the next two hours, reports and questions flew back and forth by radio and phone line about the paratroopers until, by 0530, Major Hubert Drake, the 22nd's liaison with the 4th Division headquarters was able to report that the supposed paratrooper drop was only five parachutes.

At approximately 0830, 1st and 2nd Battalions continued the attack. The activities of 2nd Battalion were described thus:

The following are quotes from the interview obtained from Capt Daniel and Capt Harrison relating to activities of July 10, "On July 10, the battalion attacked at 0830 hours with Company E and Company G. The Germans fell back to successive hedgerows, getting stronger each time. When they reached a certain point, resistance got tough as hell. Lt. DiDonato, Co E Commander, was killed by a mortar shell which hit him on the shoulder.... This was the second company commander lost by Company E in two days. Company G got pretty well shot up. No officers were left in Company E, and we sent the second in command of Company G to take charge of Company E." [27]

The Regimental Commander, Colonel Lanham, was not happy with the performance of the tank unit attached to 2nd Battalion. From 1130 to 1155, several messages were passed back and forth from 22nd Headquarters and the tank unit and eventually a liaison officer from the tank unit was sent to the 22nd Infantry to try and iron out the difficulties. The action of 2nd Battalion continued: Lt Holcomb was sent to command Company E about 1200 hours. Company F relieved Company G.

At about 1200 hours, the 3rd Battalion came around our left. That was the only thing that saved us. Company E then advanced about 500 yards. About noon, the 1st Battalion came abreast. Company F was a little to the left rear of Company E. That was the line on the night of July 10. On July 10, Private Hix got his third tank from the corner of a hedgerow behind a tree. He got three bazooka shots into the tank at less than five yards. He was so close that the explosion scorched his face. The tank crew escaped. [28]

With no officers left in Company E, the higher non-commissioned officers had to take charge, until 1st Lieutenant Hoyt Holcomb from Company G could take charge at midday. Because of the heavy German resistance, 3rd Battalion was brought up to aid in the attack and by 1345 it was on line with 2nd Battalion. Adding to the stiff resistance by the enemy, from about 1200 Edwards and his Battalion had been under fire by German artillery of a large caliber. Several messages were sent between headquarters and 2nd Battalion asking to identify the caliber of the enemy guns, but no confirmation of the size and location of the guns could be determined.

As the day wore on, Edwards' Companies were being hit hard and their effectiveness was being diminished. Edwards was having a difficult time staying in communication with his units and reports from his Companies had to be relayed to him by LTC Ruggles at headquarters. At 1610, Ruggles informed Edwards that 2nd Battalion's Company G had been hit really hard. The Company commander, LT. Jackson, had become badly unnerved. This was the same officer who earlier in the day had performed the following heroic act:

Lt. J. O. Jackson, tired of the constant harassing and dangerous fire from an enemy machine gun, climbed out of his foxhole and, before the eyes of his astonished G Company, crawled quietly around the edge of a field, over a hedgerow, pulled a pin from a hand grenade with his teeth, threw the grenade at the machine gun nest, and immediately after its explosion, rushed in to polish off the Germans with his bayonet, and then climbed back calmly over the hedgerow and back into his foxhole as he remarked succinctly, "That's the way to do it."[29]

Company F, coming up behind Company G, had to move through the shattered Company and take over its sector. To the right of Company G, Company E was also badly hit. Right at this time, Company G was attacked by enemy tanks. Lanham asked Edwards to keep a couple of the tanks attached to 2nd Battalion but turn over the majority of his tank support to 3rd Battalion, so that it could take over the attack from 2nd Battalion.

At 1712, General Barton at 4th Infantry Division headquarters spoke directly to Lanham, asking for a situation report. Lanham informed him of the tank attack against 2nd Battalion, that Company G was without officers, and that the Company commander of Company G had broken down. At 1755, Captain Burnside from 2nd Battalion told Major Kenan at headquarters that Company G had knocked out two German tanks and the enemy tank attack had been stopped. Burnside also related that 3rd Battalion had been given the majority of 2nd Battalion's tank support. Burnside told Kenan that 2nd Battalion had only two tanks left and wanted badly to keep them. He also informed Kenan that Company F had to be moved through Company G because Company G had lost all its officers. At 1833, Burnside asked Kenan to check on some friendly artillery fire that was falling short and landing in 2nd Battalion's area. At 1842, Burnside notified headquarters that 2nd Battalion was resuming the attack.

By 1900 hours, the day's advance was halted. At 1903, Edwards spoke directly to Kenan at headquarters and told him that 2nd Battalion had lost quite a few good men that day and the two officers talked over the situation. As the Regiment settled in for the night, Kenan asked if 2nd Battalion had turned over its tanks to 3rd Battalion. Edwards replied that they had been turned over. Kenan then gave Edwards the plan of attack for the next day. Leading the assault would be 3rd Battalion with 2nd Battalion in support and 1st Battalion in reserve. At 2400, the situation of the 22nd Infantry was reported to Ruggles, the Regiment's Executive Officer, as: 1st Battalion could use 11 or 12 officers, 2nd Battalion needs men and 3rd Battalion was in good shape.

Though not mentioned in the Daily Action Journal of the 22nd Infantry, Company F also lost its commander on July 10. During a German counterattack against Company F's positions that day, enemy mortar fire severely injured Lieutenant Jim Beam. During the enemy barrage, a mortar round landed close to Beam, breaking his left foot and practically severing his right foot. The unrepairable right foot was amputated on the spot by Captain Humm, 2nd Battalion's Medical Officer.

A passage from Chaplain Bill Boice's History of the 22nd illustrates the experiences of Earl Edwards and 2nd Battalion on July 10: (Note: Edwards was still a Major at the time but Boice mistakenly identifies him as a Lieutenant Colonel.)

The Battalion Commander, Lt. Colonel Earl W. Edwards, was inspecting his battalion following one of the fiercest bits of fighting on the 10th of July, reorganizing and tying in. While crossing a field from one hedgerow to another he came upon a medic, a private first class, who had been on the front lines every single hour since the battalion had landed. His Company Commander swore by him, and everybody talked of his courage. But he had had about as much as he could take. When the colonel approached, he was doing his best to aid one of those impossible tragedies of the war, a man whose skull had been laid open and who had sustained such a severe brain injury that he could not possibly live, and yet he would not die. The medic had done everything he knew how to do. He had been alone on this field for so long and the strain had been so great, and yet a devotion to duty and to a wounded member of his company was so great that he would not leave.

When he saw the Colonel, his reserve broke and he cried, "Sir, he won't die. He ought to die, but he won't. I have done everything I can for him, but he won't die. Why won't he die?" The Colonel led the aid man away, giving instructions for much needed rest and care for the boy. This was war at its lowest, purest hell. [30]

On July 11, 1944, Company E received 86 enlisted replacements. Company F received 28 enlisted replacements. Company G received 47 enlisted replacements. Company H received 10 enlisted replacements.

At 0130 on July 11, Major Kenan from headquarters spoke with Captain Burnside in 2nd Battalion and gave him the general situation and orders for 2nd Battalion for the upcoming day. The attack would commence at 0900 with 2nd Battalion moving to the left and rear of 3rd Battalion who would lead the advance. At 0840, 2nd and 3rd Battalions were notified that tank support for the attack was on the way for both of them. The attack was launched on time but intense small arms and mortar fire from the enemy caused the advance to go slowly. At 1117, Edwards reported to headquarters that he was receiving fire from the unit on their left, the 329th Infantry.

Stiff resistance brought 3rd Battalion's attack almost to a standstill. Edwards directed his Battalion to go around the left flank of 3rd Battalion in a column movement formation. At 1245, headquarters told Captain Burnside at 2nd Battalion that Colonel Lanham was not satisfied with 2nd Battalion's progress. Lanham believed that the Regiment was about to break through the enemy's lines and told Edwards to continue moving in a column formation and not to stop.

At 1415, Burnside informed headquarters that 2nd Battalion was still moving in column formation, but that Company F had been stopped by tanks and small arms fire. At 1445, Burnside told Kenan at headquarters that Company F was held up by enemy tanks and Company G was ready to break formation and go to the aid of Company F. At 1502, Burnside informed Lanham directly that Company G was moving around the left flank to help Company F. At 1504, Edwards called the artillery commander and let him know that 2nd Battalion was having trouble with its artillery support. The problem was that the artillery's liaison "spotter" planes were calling for fire on enemy tanks which had already been knocked out. Since many of these tanks were now in areas occupied by 2nd Battalion, Edwards' men were taking casualties from this fire.

Division headquarters asked regiment for a situation report at 1615 and was told that 2nd Battalion was still swinging around the flank to ease the pressure on 3rd Battalion. At 1617, Edwards informed headquarters that 2nd Battalion was receiving fire from its left flank. Edwards was told to use artillery, to which he replied that there was none there. Edwards said that his Battalion was being hit by very heavy stuff, most likely 170mm mortars. At 1630, 2nd Battalion reported the coordinates of where the enemy mortar fire was landing and the direction (azimuth) from which it was coming. Regiment received a report at 1640 that the firing from 2nd Battalion's left flank had ceased. At 1740, Edwards reported the coordinates of all of his Companies. He also reported that from radio traffic he could hear from the liaison planes, he understood that German tanks attacking 2nd Battalion were ducking under shelter at some houses.

At 1820, Edwards informed headquarters that 2nd Battalion was consolidated and getting set to push forward. At 1920, Captain Lemman in 2nd Battalion reported that Companies F and G were pushing forward. At 1950, Lemman told Major Kenan at headquarters that things were pretty rugged and the two Companies were no longer moving forward. He informed Kenan that the Companies were tied in with 3rd Battalion on the right and had sent out patrols to cover the area to the left. He told Kenan there were still some Germans in their area.

At 2040, Edwards was told by headquarters to button up for the night at 2250. Kenan told him to pay particular attention to his left flank during the night. He instructed Edwards to have Companies E and G ready for any possible enemy counterattack and to have bazookas and anti-tank guns on the flank as well. Edwards was informed that the attack would start again the next morning at 0900, trying a different tactic. The objectives would be the same as the day before, but the pre-attack artillery fire would go over 2nd Battalion's defensive lines instead of 3rd Battalion's lines. It was also suggested that for the attack, 2nd Battalion should move its anti-tank guns into the area between it and 3rd Battalion.

At 2053, Edwards was told by headquarters to button up for the night immediately instead of waiting for the previous time given. At 2155, Edwards was ordered to report to the Regimental headquarters for a briefing along with other unit commanders at 2315. At 2325, headquarters was informed by 2nd Battalion that Company F had received a German counterattack, with tanks involved, about 45 minutes previously. The attack was halted and repulsed through the use of supporting artillery fire.

Swede Henley's diary entry for July 11, 1944 was a short and curious statement about a very tough and confusing day:

11 July

Received orders to attack and seize road SE of La Mangerie (Ed., La Maugerie) — attack jumped off. Enemy counterattacked with tanks and not only held us but pinned our ears back. 2nd Bn. went into the fight on the left flank and fizzled out. Casualties were high. "I" Co. lost 47 men first — "L" Co. lost 23 men in one shelling. BN at (Ed. at = Anti-tank) platoon got 4 Mark V tanks. [31]

On the morning of July 12, the Battalions were informed that the attack would be resumed at 0915. Second and 3rd Battalions would attack as a team. At 0730, Edwards was notified that the orders assigning a platoon of tank destroyers to him were rescinded.

The attack was launched on time. At 0930, Company G reported that they were taking German artillery fire. At 1020, Company E reported that they were at the edge of the orchard along the road towards the village of Raids with Company G abreast of them. At 1048, Edwards relayed to headquarters that Company G had knocked out a machine gun nest and captured two prisoners. Edwards and his battalion held their positions at the orchard.

At 1122, he informed headquarters that 2nd Battalion was receiving artillery fire. He said that he thought it might be friendly fire coming from the 83rd Division on their flank. At 1152, Edwards gave Major Kenan at headquarters the coordinates of where the artillery fire was landing and the azimuth from which it was coming. Edwards held his Battalion in place and at 1315 reported to headquarters that his command post was located at the northwest corner of the orchard. The 83rd Division denied that it was their artillery fire impacting in the 22nd Infantry's area.

At 1437, Edwards called the Medical detachment and asked for additional medical aid for his wounded. The ferocity of the day's action was described by Chaplain Bill Boice:

The attack had only begun when Captain James B. Burnside, Second Battalion Executive Officer, was wounded, leaving the Second Battalion with only seventeen officers. The fighting became fierce hand-to-hand conflict. Casualties mounted rapidly; the rifle companies did not have over seventy effective fighting men. Though the number was small, individual courage and initiative were everywhere apparent. First Sergeant Kenyon of G Company gathered together fifteen men and took over a section of the front normally held by a platoon. He said in a rather calm undisturbed manner, "I've taken over this part of the front and I'm going to hold it. You don't need to worry about it."[32]

At 1501, Edwards told the Regimental Executive Officer, LTC

Ruggles, that an artillery forward observer had given away 2nd Battalion's position over his radio. Ruggles told Edwards to get the man's name. Edwards replied that he would deal with the increased danger caused by that mistake to the best of his ability but that he needed another Company in support. At 1522, it was reported that the front lines were still and that one platoon of Company E had been moved back to the rear of Companies F and G. Edwards informed headquarters at 1605 that 1st Battalion was moving up on his side on the road by the orchard.

The forward movement of the 22nd Infantry was halted for the day and word was passed that the 12th Infantry would be moving into the area during the night to relieve the 22nd. At 1620, headquarters instructed Edwards to send at least one man from each Company to the road junction to act as guides for the 12th Infantry as it moved up. All of those selected as guides would have to know the exact positions of all units in the area, in order to pass that information along to representatives of the 12th Infantry.

During the day, the 8th infantry, on the right flank of the 22nd Infantry's attack, had fired mortars and small arms across the front of the 22nd Infantry area. At one point, Edwards was sure some of the 8th's mortar fire had landed among his Battalion. At 1808, 2nd Battalion was told that the mortar fire which landed in their area that day was not from the 8th Infantry.

At 2215, Edward was instructed to release his tank support to the 12th Infantry who were moving in to relieve his Battalion. July 12 ended with the 22nd Infantry Regiment being relieved in their battle positions by the 12th Infantry Regiment. The attack toward the city of Perriers was over for the 22nd Infantry. From July 8 through July 12, the 22nd Infantry Regiment had suffered a total casualty count of 1,379 dead and wounded.

The following excerpts of interviews from officers of 2nd Battalion give some idea of the intensity of the fighting during the attack toward Perriers:

The following are quotes from the interview obtained from Capt. Daniel and Capt. Harrison relating to activities of July 12, "On July 12, the 2nd Battalion advanced another 300 yards. This day attack was ordered for 0915 hours. The first attack gained about 300 yards after the heaviest artillery preparation we had ever had. On July 11, Lt Beam, Commanding Company F, had his leg shot off. On July 12, Lt. James O. Jackson, Commanding Company G, was wounded. He had taken a patrol of about 15 men and was advancing along one hedgerow to another throwing hand grenades when the Germans opened fire with mortars. Lt Jackson was wounded and a total of about 50 men were casualties since the mortar fire moved on through the front lines. At 0920 hours on July 12, Capt. James B. Burnside, 2nd Battalion Executive Officer, was wounded. Two officers with Company G, three officers with Company F, one officer with Company E, and four officers with Company H were all that remained, a total of 17 officers in the battalion. In this position, the enemy suffered the heaviest casualties we have yet inflicted with small arms fire. The fighting was hand-to-hand.

"The enemy was never more than 50 or 75 yards away. The Germans all wear camouflage suits. They saw us move in and knew where we were much quicker than we could locate them. Every time a man stuck his head up over a hedgerow, the Germans would fire. In the afternoon of July 12, after the German mortar barrage, Company F was reduced to about 75 effectives, Company G had 50 and Company E about 70. (On D- Day, each of these companies had about 185 men each.) Company E, which was in reserve, sent 20 men to reinforce Company F and 50 to reinforce Company G. We alerted a company of the 12th Infantry to protect our left flank." [33]

On July 14 the following interview was recorded with Lieutenant Hoyt C. Holcomb, Executive Officer of Company G who by then was commanding Company E:

"Scouts are always out on the field ahead of us; two for each platoon. It takes one company to take a normal size field and two companies to take a large field. We made two attacks on the main line July 11 and one on July 12. On July 12, Company G reached the crossroads, but Company F was still held up in the orchard. The Germans opened a violent mortar barrage, and survivors of Company G withdrew.

"I estimate the Germans had eight or 10 tanks along the roads behind their positions. They had one tank in particular moving up and down the road in their position sounding its siren. It was probably an effort to impress their men.

"There was no way to tell which hedgerows, or which side, the Germans will be on. They may be anywhere; hence, no way to tell which approach (if any) is covered. They usually have machine guns and mortars sited in corners to sweep the length of hedgerows, also to fire across and sweep the open fields. Although we expect German guns to sweep the hedges, we still prefer to move along the hedgerows. We never use skirmish lines. The best formation is column. The columns advance along hedgerows on both sides of a field and several adjoining fields. The first to reach the next lateral hedgerow will flank out the enemy and assist the advance of others." [34]

On July 14, the following interview was recorded with First Sgt. William Kenyon, Company G:

"Tanks in the sunken road moved up and down their side of the hedgerows. The Germans used the tanks by moving them back and forth on the sunken road firing 88's at our personnel. They continually changed positions. They would sometimes move around to the houses on the left flank. In the last attack, we used seven medium tanks and one light tank and a bulldozer. Two were knocked out and the remainder withdrew. The enemy, on the last day, was showing signs of weakness. We took four prisoners in Company G and the Germans were not picking up their wounded. I think we could have advanced if we had more men.

"Company G has only five non-coms left. Four of them are shock cases and it has no old officers. This lack of leaders is the greatest handicap. The new men, who have never been under fire, freeze in their holes when the firing starts. The company commander goes around and boots them out, only to have them hide in the holes again as soon as he is past. After the tremendous barrage placed on the German positions on July 12, the attack troops did not see any evidence that many Germans had been killed. We cannot estimate the number of German dead since the Germans removed them at night but, in this case, attacking right behind the barrage, they would have seen the slaughter if there had been one."[35]

First Sergeant William L. Kenyon of Company G mentioned above would never return to the United States. He would be killed while still serving in Company G, 2nd Battalion during the 22nd Infantry's attack against the city of Prüm, Germany on February 8, 1945.

On July 13, 1944 Headquarters Company 2nd Battalion received six enlisted replacements. Company E received 46 enlisted replacements. Company F received 46 enlisted replacements. Company G received 52 enlisted replacements. Company H received six enlisted replacements.

By 0250 on the morning of July 13, all of 2nd Battalion except for one platoon had withdrawn from the front lines and headed back to the rear area. The regiment was being held in reserve by the

4th Infantry Division near the village of Les Forges. By 0300, all of 2nd Battalion was in the rear area. Edwards and his Battalion, along with the rest of the 22nd Infantry, had thus ended their part in the campaign of the hedgerows of Normandy.

Major Earl Edwards was promoted to Lieutenant Colonel (AUS) on July 13, 1944. At 1437 on July 13, Major Hubert Drake assumed command of 2nd Battalion. Edwards moved to the Regimental Staff of the 22nd Infantry as S-3 Operations Officer, replacing LTC Thomas Kenan who had been wounded.

As Operations Officer, Edwards was the direct contact between the 22nd Infantry and the 2nd Armored Division in the upcoming Operation Cobra breakout from Normandy. Elements of the 22nd Infantry began training and rehearsing for the upcoming operation with elements of the 66th Armored Regiment of the 2nd Armored Division. On July 19, the 22nd Infantry Regiment was officially subordinated to the command of Combat Command "A" of the 2nd Armored Division. Combat Command "A" was commanded by Brigadier General Maurice Rose. Edwards was officially assigned as the Liaison Officer between the 2nd Armored Division and the 22nd Infantry Regiment.

The radio logs of the 22nd Infantry during the time period of Operation Cobra show that Edwards was constantly communicating with all elements of the 22nd Infantry and the command echelons of the 66th Armored Regiment, the 2nd Armored Division and the 4th Infantry Division. He directed all movements of the 22nd Infantry Regiment in the operation and was the central figure to whom all units in the 22nd Infantry reported. Edwards was in charge of getting supplies to the rifle companies, solving their communication problems, directing other units such as Engineers, Cannon Company, Anti-tank Company, and artillery to furnish support for the 22nd Infantry Battalions and individual Companies where necessary, monitoring and directing movements of trucks as transportation for the rifle companies, and any and all items of command and control needed during the attack.

Colonel Lanham commanded the 22nd Infantry Regiment and during the operation he acted as the Infantry Advisor on General Maurice Rose's staff. However, during Operation Cobra it was Lieutenant Colonel Earl Edwards who directed all activities of the 22nd Infantry Regiment. Edwards was also assigned to General Rose's staff and related all commands and directives coming from Lanham or Combat Command "A" to all units in the 22nd Infantry. Edwards coordinated all movements of the 22nd Infantry and ensured the union of the Infantry with Armor was successful. The radio logs show that Edwards got very little sleep during the operation as he was on the radio or phone well past midnight of each day.

The following story concerning Edwards during Operation Cobra was written by Don Warner, who, as a young officer with the 22nd Infantry Regiment rose from Platoon Leader to Company Commander to finally Battalion S-3 Officer for 3rd Battalion. The story is written "tongue in cheek" and Warner is kidding when he says his friend Lum Edwards had very little to do during the operation.

"During the St. Lo breakthrough in late July 1944, Lum Edwards inherited a man by the name of Stone as his driver. As part of the combined Infantry-tank team spearheading the breakout through the German lines, Edwards, as Regimental S-3, had very little to do and was assigned to the staff of General Rose of the 2nd Armored Division. He was to act as liaison between the 22nd and the 2nd Armored as needed. He had no idea that General Rose always acted as his own scout and was always where the action was the greatest. Lum had no choice but to accompany General Rose wherever he went on the battlefield. This meant that his driver, Stone, followed Lum.

"In so doing, in an open space in a barn yard, Stone noticed a horse and cow watering trough full of water. Stone then told LTC Edwards that he thought the Colonel REALLY needed a bath! Edwards was momentarily shocked to think that an enlisted man would make such a suggestion but knew that Stone was right. Edwards was also quick to point out that he religiously followed COL Buck Lanham's instructions on field hygiene in that he shaved every day, washed his feet, his family jewels, and under his arms. Stone still insisted that the Colonel really needed a bath and needed it real bad.

"Lum was shown the water trough out in the open barn yard. Lum told Stone that he didn't want to strip down in front of all the soldiers who had set up defensive positions in the stone buildings surrounding his bath trough and were watching every move. Stone told Lum to go ahead, strip down and get in and soak, relax, and enjoy his bath because he REALLY needed it.

"Lum, at last, followed Stone's instructions, stripped off and entered the green moss filled water trough to enjoy the luxury of a bath. Stone told the others to look the other way while Lum reverted to his birthday suit and took the plunge. The men cheered and clapped. Stone told Lum to stay in, soak and enjoy the bath—he and the other men would be on the lookout for any Germans.

"About this time, the Germans came out of a draw, in force, and attacked this location. Small arms fire covered the entire area, and it was nip and tuck for thirty minutes with Lum pinned down inside the water trough. All during the period, Stone kept calling to Lum to keep his head down and enjoy his soak and that he was trying to arrange a counterattack to get him out. Throughout the counterattack, Lum could constantly hear Stone's voice telling him to enjoy his bath, they would get him out.

"When the water trough was recaptured and Lum surfaced, he was blue from exposure, covered with moss and trembling from cold, although it was a hot July day. Stone spoke the first words following Lum's rescue, "Colonel, I know you enjoyed your soak, you really needed it."

"Perhaps Lum Edwards should be recommended for the British Order of the Bath with moss cluster."[36]

Operation Cobra ended on July 31 and on August 2, the 22nd Infantry was relieved of its assignment to the 2nd Armored Division and reverted back to command by the 4th Infantry Division. General Rose issued a letter of commendation to the 22nd Infantry in which he stated he had never worked with a finer Infantry unit. For its actions and performance in Operation Cobra the 22nd Infantry Regiment was awarded the Presidential Unit Citation.

As the 22nd Infantry moved across France and Belgium in its attack toward Germany, Edwards continued his duties as Operations Officer. He was responsible for drawing up all orders of movement and plans of attack. He and his Commanding Officer, Colonel Charles T. "Buck" Lanham, then discussed the merits of all such orders and plans. Once they were in agreement, it was up to Edwards to issue the finalized orders to the various units in the Regiment and to see to it that the orders were carried out. It was Edwards' job to keep the Regiment running smoothly. In his later years, Edwards wrote that he was seldom less than a dozen yards away from Lanham at any given time from July 1944 through March 1945.

On July 29, 1944, the famous author and correspondent Ernest Hemingway attached himself to the 22nd Infantry Regiment.

Hemingway and another correspondent found the 22nd Infantry command post in a small farmhouse near Le Mesnil-Herman:

Lieut. Col. E. W. "Lum" Edwards, operations officer, was "awfully busy" when the tall, grizzled man and his shorter companion appeared and asked to see Col. Charles Trueman Lanham, who was equally busy in the front room of the farmhouse where his operations maps were posted...Lum Edwards told him of the visitors. One was a war correspondent; the other, he understood, was a Colonel Colliers from Washington..."Colonel Colliers?" said he, cocking his head and holding out his hand. "I'm no colonel," the visitor said. "I'm a correspondent for Collier's. My name is Hemingway." "Ernest, no doubt," said Lanham. "Yes, my name is Ernest." [36a]

Hemingway remained mostly with the Regimental staff and therefore became personal friends with Colonel Lanham and Earl Edwards. The friendship continued on after the war.

August saw the 22nd Infantry racing across France, moving through Paris on August 27, through Belgium shortly thereafter, and reaching the border with Germany on September 11.

Hemingway was known for his partying and grand behavior and lavish dinners. On the night when the 22nd Infantry first penetrated Germany, September 12, 1944, Hemingway held such a dinner and Lanham and his Regimental Officers (including Edwards) and Battalion Commanders attended. During the dinner, German artillery began falling in the courtyard outside the house in which the dinner was being held. Shrapnel perforated the windows of the house, but no one was injured.

On September 16, 1944, Edwards lost his good friend, Lieutenant Colonel John Dowdy, who was killed by German artillery just outside of Sellerich. Later in life when Edwards wrote his memoirs he recalled the day:

"John had just called me on the phone (I was now S-3 of the regiment) and said he needed to see me about a problem he had and was on the way to the regimental Command Post. I had walked down to the entrance waiting for him when I received word he had been killed. It was a blow, to me. I had known his mother well. She doted on John, he was her life, and she was a very disturbed woman thereafter. She never was able to deal with his death. She worried over whether to bring his body home after the war—he did, and I attended his funeral, along with some of his best friends. One time Mrs. Dowdy visited Mother in Cruger to try to talk out her problem about her son's death with her. This is, perhaps, a typical example of the kinds of tragedies a war brings that few ever know anything about."[37]

Lanham's plywood trailer, the scene of many talks between Ernest and the staff of the 22nd, was a masterpiece of rolling architecture, drawn from place to place by weapons carriers or 2 1/2-ton trucks. It contained two bunks, a stove, a drop-leaf table, a washstand, two settees, and a field telephone.[37a]

Edwards would not stay in the command trailer overnight. He felt the trailer was an invitation to enemy fire. No matter how bad the weather conditions were, he would sleep in a tent or foxhole and always refused to sleep in the trailer.

During the rest of September and the beginning of October, the 22nd Infantry Regiment continued operations in Germany in the area around Brandscheid and Hontheim. In early October, the Regiment's activities were shifted a bit to the north in the area around Bullingen. In early November, the Regiment moved to Zweifall, Germany to prepare for the attack in the Hürtgen Forest.

The 22nd Infantry Regiment began its attack in the Hürtgen Forest on November 16, 1944. The command post trailer could not be used in the heavily wooded Forest. Colonel Lanham and his Regimental Staff, including Lieutenant Colonel Earl "Lum" Edwards, commanded the Regiment from muddy foxholes amid the densely packed trees from November 19 to December 3.

Donald A. Warner, who started the attack in the Hürtgen as a Platoon Leader in Company A and ended the attack commanding the Company, became a life time friend of Earl Edwards. Many years after the war, Warner wrote a letter to John King, the nephew of Earl Edwards, in which he described the activities of his friend "Lum" Edwards during the Battle of the Hürtgen Forest:

"In the 22nd, Lanham took complete charge during the day. During the darkness, Lt/Col Ruggles was in complete charge. Your over worked uncle was available to both day & night, 24 hours around the clock. (As his assistant, I wondered if he ever "hit the sack"? Sleep)

"In the Hurtgen, it was dark, dark, dark around 1630 & cold, snow, mud etc. John R. Ruggles & Lum Edwards were in Lum's foxhole, with a wet wool GI blanket, a large map board, a GI right angle flash light. Lum's foxhole had both radio and sound power phone hooked up to all battalions.

"He, Lum, would discuss the division order, supporting units, objectives, plan of attack, reserves, along with a million other items. Nothing overlooked.

"Plans, when jelled, were written up and hand carried to each of the three battalion HQs by liaison officers. (How in the devil they accomplished their mission I'll never know. To me, they were the unsung heroes).

"Back on course—In drawing up plans for the morning jump off, a shell landed just in front of Lum's CP. Both he and Ruggles were squeezed shoulder to shoulder, constant in coming kraut artillery fire around the clock, both under the wet GI blanket with flash light and map board when a 120mm shell landed just in front of the CP. Shrapnel in all directions with a fist size chunk traveling across the foxhole, front to rear, passing through and completely destroying the map board & map and cutting down a four inch tree just to the rear of the CP. In the fist size chunk of shrapnel, avenue of travel at a high rate of speed, its route was between the heads of Lum & Ruggles, a space of not over four inches. Both should have been on the way to the happy hunting grounds." [38]

On December 3, 1944, the 22nd Infantry was relieved from the attack in the Hürtgen and sent to Luxembourg for rest and refit. On December 16, the Germans attacked through the Ardennes, thus beginning the Battle of the Bulge. The 22nd Infantry Regiment as part of the 4th Infantry Division held the southernmost edge of the Bulge. By the end of January 1945, the German attack had been stopped and the 22nd Infantry Regiment was moved to northeast of Bastogne, Belgium. S-3 Officer Earl Edwards drew up plans for the Regiment's next assignment, an attack into Germany in the same place it had attacked back in September.

Captain Don Warner was now assigned as Edwards' assistant. In a letter written to Edwards' nephew John King long after the war, Warner described what it was like to be Edwards' assistant:

"Following the Battle of The Bulge when as you British say, "the bits & pieces settle down", I was transferred to regiment as assistant S-3 or operations officer, Col Edwards assistant.

"Fighting a company with a jump to Regiment was truly a different ball game. Col. Edwards zeroed me in and advised me to ask questions, etc. He was truly a master at his task, which greatly eased my new burden. Orders out to all battalions complete with maps, etc.

"Together with halftrack driver, we followed all attacks, giving full coordination to all units & supporting units. This halftrack was our rolling command post. Since several night attacks were made, this halftrack was blacked out with a tarpaulin. Col. Edwards rode in the front right seat, keeping in full radio contact with attacking units and with Regimental HQs. As he repeated map coordinates and other information to me, all information was plotted on the map board. Don't forget we each only had a GI flashlight held in one hand with the other either talking to attacking units or HQs.

"Yes, it was exciting to plot advancing fighting units' progress, plotting same on the map board as Col. Edwards kept in touch with each by radio net. At times, he reviewed map plottings and recommended minor changes to various units until the objective was taken. (Most Army units used the term GOOSE EGG and not OBJECTIVE)

"After front line units were dug in for the night, orders sent for next day's JUMP OFF, night combat & reconnaissance patrols sent out, Col. Edwards, Sgt. Ulm, and I would sit down, review all messages written, along with radio logs, maps used that day, etc., and write up the Regiment's ACTION AFTER ACTION REPORT for the day. This was most trying for each of us being so late at night with first light at 0500 Hour which was an hour or two away. Sleep? What's that? We did cat-nap from time to time, but always on post & with one eye open.

In later years, the War Department or Department of Defense said that the 22nd had the best kept records of all other units in the ETO. Thanks to Col. Edwards.

"John, the above is a once over, touching briefly on my association with your uncle. He was a real leader and taught me well." [39]

During the month of January, 1945, the 22nd Infantry Regiment was headquartered in Luxembourg and Belgium, operating along the German border. On February 3, Edwards' attack plans were implemented, and the Regiment attacked toward Brandscheid.

Brandscheid was taken and the attack continued against the

city of Prüm. The fighting was intense, and casualties in the 22nd Infantry were heavy, but Prüm was taken by mid-February. For the next month, the Regiment pushed deeper into Germany.

On March 3, 1945, Colonel Charles T. Lanham was promoted to Assistant Commander of the 104th Infantry Division and Lieutenant Colonel John Ruggles assumed command of the 22nd Infantry Regiment. Lieutenant Colonel Arthur Teague became Executive Officer, and Lieutenant Colonel Earl Edwards remained as S-3 Operations Officer.

Near the end of March, the 22nd Infantry Regiment attack into Germany turned toward the south, and in early April, the Regiment had taken Bad Mergentheim. Crossing the Danube River in late April, the 22nd Infantry Regiment moved further into Bavaria toward the Alps to the Isar River.

The Regiment was relieved of front line duty on May 3, 1945 and moved to the vicinity of Nürnberg (Nuremberg), Germany to begin occupation duty.

After the cessation of hostilities, and until the early part of June, the 22nd Infantry Regiment occupied its assigned occupational area of Landkreis Ansbach, Landkreis Feuchtwangen, and Landkreis Dinkelsbuhl, guarded military installations, operated Displaced Persons' Camps, processed prisoners-of-war, patrolled by motor an area of approximately 1,400 square kilometers, enforced military government regulations, and maintained military control of the sector. Men of the Regiment fretted under non-fraternization regulations and prepared themselves for what they presumed to be an extended tour of occupation duty.

Early in June, however, word was received that the Regiment would return to the States, pause briefly for 30 days of rest and recuperation, and proceed to the Pacific, there to engage in further operations. [40]

Edwards recalled the time when he wrote his memoirs later in life. He began with the days immediately following May 3:

"The 22nd Infantry was just in the edge of the Alps near the Austrian border at this time. There is no way to describe my feelings when we received a message to cease all combat operations at a certain hour, so I won't try. The Fourth Infantry Division Commanding General came by and instructed us to proceed to the vicinity of Nurenberg, Germany and occupy an assigned sector. We were told to prepare to be in Germany for three to five years as an occupying force.

"We promptly moved to our new location and set up camp. Regimental Headquarters was located in a small town named Schwarbach. Col. Ruggles, now our Regimental Commander, and Art Teague, the Executive Officer, and I selected a very small but very nice house near the headquarters to live in. Life was good again.

This was a time of considerable confusion — the Army set up a system of points to decide who was to be sent home first. Most of the draftees were anxious to get home as soon as possible, so there was a lot of counting points. Every week or so we would have a small ceremony seeing some of the men off — back to the USA and civilian life — a happy occasion, to say the least.

Our sector had some Prisoner of War camps located in it, as well as a few large refugee camps of Russian or Polish nationals. We assigned officers to oversee and administer these facilities.

Col. Ruggles and I would visit the various battalion locations to see that everything was in order and that, at least, a minimum program of some sort was being carried on. These were lazy, restful times for us with very little pressure from high headquarters.

In time, I was required to organize truck convoys to transfer the refugees back to Russian control. I made several trips in long truck convoys from Ansbach, Germany to Pilsen, Czechoslovakia for this purpose. Time passed slowly but pleasantly.

The plan for us to remain in Germany on occupation duty for three to five years was abruptly canceled in June, 1945. It seemed that several divisions had been selected to be sent to the Pacific to be a part of the planned invasion of Japan, and alas!!, we were one of them. The good part was we were to proceed to Japan by way of the U.S.

Most of the regiment was sent to the French port of Le Havre by rail, but the vehicles were required to go overland. Col. Ruggles put me in command of this operation, and he rode along with us. It took us three days to reach Le Havre. Soon after our arrival, we were loaded on a troop transport and on July 2nd, 1945, we headed for the U.S." [41]

Except for Companies H and M, the 22nd Infantry returned to the United States aboard the *United States Army Transport (U.S.A.T.) James Parker* on July 3, 1945. The *Parker* had been built before the war as a passenger ship and was requisitioned by the Army and converted to a troop transport in 1941. Captain Don Warner described the voyage home aboard the Parker:

"The crossing weather wise was smooth and on deck was beyond description. All had regular beds, kitchen open 24 hours a day, decks loaded with crates of fresh fruits, lettuce, and frozen quarts of milk. One could enter the dining room, be seated at a table with beautiful place settings, given a menu full of outstanding selections from A to Z, order a drink or three, sit back, and leisurely enjoy a feast. The waiters were ship crew and truly outstanding. Following this meal, one returned to his berth, complete with clean sheets, and enjoyed a siesta.

"In mid ocean on the 4th of July, the entire ship's crew went on strike and all services came to an abrupt halt. In general, all crews from the captain on down were union members making big bucks, with many perks over and above, such as danger pay, etc.

"A little meeting, off the records, was called by the Regimental staff asking all present how soon we would like to arrive home? Col. Ruggles said that all men in the 22nd still had their weapons didn't they? Meeting adjourned!

"The ship was taken over from top to bottom and once more we were under way to NY, NY. We made it by taking over the entire crew and ship. Upon entering NY harbor and passing the Statue of Liberty, one could not help but notice the circling welcome home ships complete with blaring brass bands playing their hearts out for us.

"Upon docking, the ship captain with his four mates departed first, making a phone call to his union and the Coast Guard, regarding the July 4 high seas episode. They did not get to first base, period." [42]

The *Parker* docked in New York Harbor and all soldiers proceeded to Camp Kilmer, New Jersey. The entire Regiment was authorized 30 days leave, beginning immediately. Men were arranged in geographic groups and boarded trains bringing them to Army installations close to their homes. Lieutenant Colonel Earl Edwards was placed in charge of 102 men from Mississippi and Louisiana whose destination was to be Camp Shelby, Mississippi. [43]

The journey to Camp Shelby took three days because of heavy railroad traffic at that time. The train had to go to Mississippi by way of Chicago, Illinois. Don Warner remembered the trip in a letter he wrote to John King, nephew of Earl Edwards:

"For some unknown reason, our train was delayed until the next day prior to departure. The sun was hot, and all windows were up. At several train stations along the way, we were switched to a side-track in order for regular trains to keep schedule.

"One early afternoon on a side track in front of a depot in either Iowa or Indiana, weather hot as blazes, a young white boy walked by enjoying a frozen snow cone that caught Col. Edwards eye. Our car had eight officers and Lum called the boy over, asking him if he could purchase eight snow cones for us and one for himself? The young boy replied, "No problem, I do this for all troop trains." Lum gave him a $20.00 bill and told him to hurry. The boy took off. Fifteen minutes later, the train blew its whistle and departed. No boy, no snow cones, and no change." [44]

In his memoirs Earl Edwards continued the story:

"When I got off the train at Camp Shelby, I was met by Junior Griffin from Cruger and spent the night with him and his wife. The next day I boarded the bus for Cruger.

"It was great to be back home safe and sound after such a long and eventful time. A lot of water had gone under the bridge during that interval.

"During this leave time, word came over the radio that a huge bomb had been dropped by the U. S. Air Force on Japan—it was the atomic bomb, but we had no idea at that time. The construction of an atomic bomb was one of the best kept secrets of the war—for Americans, at least. We later learned that, as a result of espionage, the Russians knew all about it. Shortly after that, the Japanese surrendered. I can remember hearing this news on the radio and what a relief it was. This meant the entire World War II was at an end and I had made it through.

"By this time, 1945, ten years after my graduation from high school, all the girls I knew were married and gone. Someone told me that Nan Lowe had not married and was still at home. She had been engaged to marry an Air Force pilot but shortly before the marriage was scheduled, he was killed in a crash. I had known her slightly before the war, so I called her for a date. We dated several times, and, in the process, I met her younger sister, Ree. This was it. Lightning struck and I knew this was the girl I had been looking for all this time.

"We dated a few times before my leave was over and I was ordered to return to the Regiment now located at Camp Butner, North Carolina. This was August of 1945.

"These were turbulent times in the AArmy. With both the European War and the Japanese War ended, it was clear that the Army would be revamped and drastically reduced. Most of the non-regular officers and enlisted men were just waiting around to be discharged. I was assigned as S-3 on my return, as usual, but in January, 1946, I was reassigned to my old post as 2nd Battalion Commander. We spent our time with routine training and busy work. I lived in a BOQ with all the unmarried officers. Not surprisingly, some of my bachelor friends returned from our 30-days leave married. This reduced the unmarried ranks somewhat; Autrey Maroun, who had been transferred out of the Regiment before we went overseas, had been reassigned. He and I resided in the BOQ and became close friends. Almost every night we would walk up to the Post Theater and attend a movie. What else was there to do? We didn't have cars and town was a ways off.

"I was assigned as President of a Court Martial and since we had a huge backlog of pent-up cases, I spent a long time in this distasteful activity.

"At one time, I had orders, along with Tom Kenan, to attend a British Staff College located in Bombay, India. This order was soon cancelled.

"To our amazement and great disappointment, the Department of Army had decided to inactivate the 22nd Infantry Regiment and 4th Infantry Division. We simply couldn't believe it.

"This order was hard for all of us to accept and all of us made known our dismay by telephone calls and letters to anyone we thought might be able to turn this situation around.

"All to no avail. We were duly disbanded. Regular Army personnel were assigned to other units and non-regular personnel were discharged. It was an emotional time for all of us. I had served in this Regiment from July 1, 1940 until now (March, 1946) and had been promoted from second Lieutenant to Lt. Col…all in the same Regiment. It had become a home to me. It was sad to realize that after all we had been through together that this was the end of it." [45]

On February 20, 1946, Colonel John Ruggles took leave prior to attending the Command and General Staff School, leaving the 22nd Infantry Regiment under the command of Lieutenant Colonel Arthur Teague. It was Teague's sad duty to inactivate the Regiment.

As his final official act on behalf of the Regiment, Earl Edwards brought the Colors of the 22nd Infantry to Washington, D.C. where he presented them to the Department of the Army to be inactivated and stored. [46]

In his memoirs, Edwards ended his days with the 22nd Infantry:

"Just before we departed, Chaplain Boice had a great idea. He called a meeting of all of the officers in the chapel and proposed that we form an association so that we could have a way of keeping in touch with each other. We did and this organization, greatly expanded, exists until this day as the 22nd Infantry Regiment Society. We had a newsletter and have regular conventions every two years. In this way, we all get to see and keep up with each other." [47]

Don Warner wrote of Earl "Lum" Edwards:

"Lum was a master of any jobs and had perfection. He could handle all staff duties such as S-1, S-2, S-3, and S-4. In addition, his qualifications as battalion commander ranked second to none." [48]

Colonel Charles T. Lanham in a letter to Ernest Hemingway wrote of Earl Edwards:

"He was a hellish good 3 (S-3)—you and I both know that. Also spoke his mind and fought like hell for battalion commanders, which I liked plenty. Also was voice of caution and sort of official conscience—which was very good for an impetuous gent like me to have around." [49]

Earl Edwards as a Lieutenant in the 22nd Infantry. 22nd Infantry officer's brass on his coat lapel and 22nd Infantry Distinctive Unit Insignia on his shoulder strap.

John Dowdy, Fellow officer and good friend of Earl Edwards. Dowdy would be killed on September 16, 1944 during the 22nd Infantry's first penetration into Germany.

Above: The heading of a Company photo for Company B 22nd Infantry at Fort Benning 1941. 2nd Lieutenant Earl W. Edwards is the top photo as Commanding Officer which dates the photo as having been taken prior to October 1941.

2nd Lieutenant Earl W. Edwards as Commanding Officer
Company B 22nd Infantry at Fort Benning, Georgia 1941

Glenn D. Walker. Seen here as a Major General in Command of
the 4th Infantry Division 1969-1970 in Vietnam. Walker served
in the 22nd Infantry from 1941 until wounded in the Hürtgen
Forest in 1944. Edwards and Walker became lifelong friends.

Part of the Fifteenth Command and General Staff Class
at Fort Leavenworth, Kansas , September-November 1943.
Major Earl Edwards is in the front row on the far left.

Earl Edwards as a Major. Photo taken prior to D-Day

The leaders of the 22nd Infantry Regiment in the marshalling area in England just before loading into the ships to make the assault across the Channel. Earl Edwards as Commander of 2nd Battalion is standing in the rear row on the far right.

Major Earl Edwards looks out over the battlefield
from a captured German position.

James B. Burnside. Executive Officer 2nd Battalion 22nd Infantry
Regiment Wounded on July 12, he would return to the Regiment
and end the war as a Major and S-4 Regimental Supply Officer

Captain Joseph C. Rickerhauser, Assistant S-3 Officer to Lieutenant Colonel Earl Edwards beginning on July 23, 1944. Rickerhauser was from Union, New Jersey.

Michael D. Belis

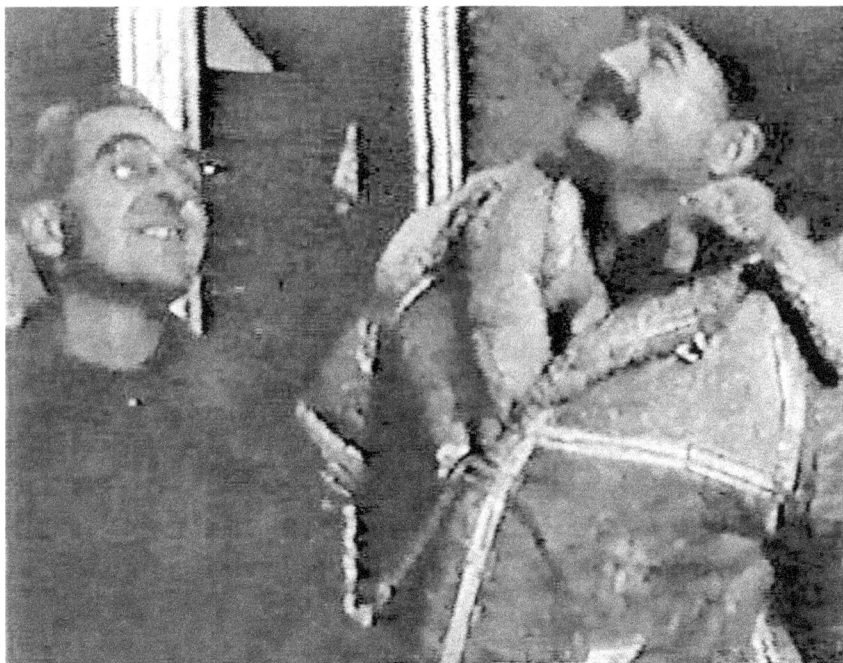

Left, Colonel Buck Lanham—Right, Ernest
Hemingway in Germany September 1944

The command post trailer used by Colonel Buck Lanham and Lieutenant
Colonel Earl Edwards as they raced across France, Belgium and into
Germany. They would be joined in the trailer by Ernest Hemingway.

The building in Zweifall used as the command post (CP) of the
22nd Infantry Regiment from November 9-19, 1944.

Edwards' command post in the Hürtgen Forest was among the trees
on top of the hill in the above photo. It overlooked a lake at the edge
of the area of responsibility of the 22nd Infantry Regiment. Across
the lake was the area occupied by the 8th Infantry Regiment.

The view looking out from Edwards' command post across the lake toward the area of responsibility of the 8th Infantry Regiment.

Left to right: Lieutenant Colonel George M. Goforth Commanding Officer 1st Battalion 22nd Infantry, Lieutenant Colonel Earl W. Edwards S-3 Operations Officer 22nd Infantry, Lieutenant Colonel John F. Ruggles Executive Officer 22nd Infantry Regiment.

On the left, LTC Earl Edwards. On the right, MAJ Tommy Harrison.
With a liberated German car which has been repainted olive drab
and marked with white stars denoting it as a U.S. Army vehicle.

Soldbuch brought home as a souvenir by Earl Edwards. The Soldbuch was the personal identification book of a Germansoldier.

This Soldbuch belonged to Private First Class Rudolf Sedlaczek

7. <u>Results of Operations</u>: Combat Team 22 attacked, reduced resistance at the cross-roads at coordinates (995822); captured the SIEGFRIED strongpoint of BRANDSCHEID; and pushed a defensive flank eastward to hill (coordinates 007820). The Combat Team was prepared for the relief of the 3rd Battalion in BRANDSCHEID and an attack southeastward toward PRUM on 6 February. Casualties were light; and 93 prisoners were taken.

LANHAM, CO

OFFICIAL:

Edwards jr
EDWARDS, S-3.

Annex:
1 - Ovly
2 - FO #3.

The S-3 Periodic Report No. 202 of Combat Team 22 dated February 5, 1945 signed in the lower left hand corner by Lieutenant Colonel Earl Edwards as S-3 Officer.

On the far left is LTC Earl Edwards at Bad
Mergentheim, Germany, April 1945

Lieutenant Colonel Thomas A. Kenan Graduating from the Citadel in 1939, Kenan was assigned to the 22nd Infantry Regiment that year as a 2nd Lieutenant. He rose to Company Commander and was the S-3 Officer of the Regiment on D-Day until wounded in July and replaced by Earl Edwards. Kenan returned to the 22nd Infantry and led 2nd Battalion in the Hürtgen Forest. He ended the war as the Assistant Executive Officer of the Regiment.

EARL W. EDWARDS
Lieutenant Colonel
S-3
Cruger, Miss.

Earl W. Edwards from the 1946 22nd Infantry yearbook

3. Post World War II Service

In his memoirs, Earl Edwards wrote about his life with his wife, Ree, and his Army service from 1946 to July 1955.

At that point in time his memoirs abruptly end.

This section will reproduce his memoir of that 1946 to 1955 time period in their entirety, as it is felt that no one can tell that part of his story as well as he can.

The author of this presentation does not have access to all of Earl W. Edwards' service records. Therefore, using Edwards' memoirs and what scant documents of his service record that are available, the following narrative of his postwar service is offered.

Edwards still officially held the rank of 1st Lieutenant in the Regular Army but did not revert back to that rank with the end of the war. Since he was Regular Army, he kept his AUS temporary rank of Lieutenant Colonel. Though not mentioned by name in his memoirs, Edwards attended two short Army classes sometime in 1946 which are listed in his service record. He attended a one week R.O.T.C. Orientation course and also a three week Professor of Military Science and Tactics course. In his memoirs, Earl

Edwards continued his story after the deactivation of the 22nd Infantry:

Realizing that I was to be reassigned and wishing to keep in contact with Ree, I called a buddy I knew in the Pentagon Assignment Office. He had been a Lt. in the old 11th Infantry. I asked him to assign me to ROTC duty with Gulf Coast Military Academy located at Gulfport, Mississippi. This I considered a fine strategic location since Ree was teaching school at the time in Columbia, Mississippi. Besides, this would put me back in Mississippi and within driving distance for occasional visits home. He said, "No problem", and on March 5, 1946, I was temporarily assigned to Ft. McClellan, awaiting orders.

I had only been there about three weeks, living in a tent again, when I received orders to ROTC at the University of Florida in Gainesville, Florida. What a shock! So I indignantly got on the phone for an explanation…what kind of buddy would do a thing like that to a friend?

He explained: The Department of Army was conducting a nationwide conference on the future of ROTC. The President of the University of Florida was a Dr. Tigert. He was also President of the Association of Land Grant Colleges that conducted the ROTC programs. At the end of the meeting, he had requested to be allowed to select the officers to be assigned to Florida. The Army Chief of Staff had approved the request so he had been conducted to the office that made the assignments so he could review the records of officers available. It just happened that my buddy had placed my record in the stack to be assigned. Would you believe that out of this stack he asked for me to be assigned to Florida. I guess that is an example of what Robert Burns meant when he wrote that often "the best laid plans of mice and men gang aft agley."

There was nothing I could do about it so off to Florida I went.

THE FLORIDA YEARS

The University of Florida is located in Gainesville, Florida, then a sleepy little college town in north central Florida. The University seemed to be its principal business. Life there was about as peaceful and quiet as it can get. This suited me just fine at this time in my life. The AArmy would obviously be in a state of turmoil as it down-sized from its huge wartime level to its much smaller peacetime status. What better place to sit this process out than on ROTC duty in a quiet little backwater town like Gainesville?

I arrived in April 1946 and reported in to a crusty old Artillery Colonel named Edmonson. He set me straight right off as to how things were going to be. He told me about reporting to his first post, an Artillery Battery in Oklahoma. His Battery Commander informed him that he had always felt that all extra duties, such as stables, duty, etc., should be shared fifty-fifty. "I've done them the last thirty years," he said, "so you should do them for the next thirty." In this vein he explained that this was his last post since he would retire soon and, of course, he would be very busy getting ready. I was to be head of the Infantry Section, and, in addition, I would be Executive Officer. In this capacity he expected me to oversee the day to day operations of the entire department. He said, of course, he would drop by as often as necessary to make any major decisions. He was as good as his word. He even had me assigned in his place on the Faculty Council as well as the Faculty Disciplinary Board. He and I got along just fine.

I soon met a Major Ernie Lorenz, head of the Artillery Section, and a Lt. Col. Joe Gillespie, head of the Air Force Section.

Joe was married and was buying a house to live in but Ernie was a bachelor, so he and I hooked up to find us a room for rent. It had to be near the campus since we didn't have cars. (I had sold mine during the war to Mr. J. T. Parker to run his mail route in.) We soon found one in the home of a Mrs. McCullar...it rented for $30.00 which came to $15.00 each. The price pleased us very much; we were penny pinchers of the first order, being children of the GREAT DEPRESSION.

We found a Mr. Leo Lemmerman already living there. He was a science professor at the University working on his doctor's degree. More about him later. We got along fine, enjoying each other's company, going out to supper each night together.

Ernie and I worked well together, but our non-professional interest had little in common, so Joe, who turned out to be a real sport, and I became the best of friends. We spent many a day hunting, fishing, and going to ball games together. Later in our careers when we were both considering retirement, Joe proposed that he and I go into business together. It was tempting but I simply didn't have the capitol to be an equal partner and, too, Ree and I wanted to get back to Mississippi to be with our families.

We had an excellent contingent of young officers and enlisted men, so my duties were most pleasant and undemanding. I soon settled into a routine of teaching a few classes each day and coordinating the several department activities, along with routine administrative duties.

I quickly discovered there was much to recommend Gainesville as a place to live, quite aside from its quite peaceful atmosphere. Fishing was great with two famous lakes, Orange, and Santa Fe, being nearby as well as several other excellent, smaller lakes handy. The quail population was good, so I quickly purchased a new shotgun to take full advantage of this, my favorite sport. (I reasoned it would be smart to make this purchase now before I was married rather than after.) The Gulf of Mexico and the Atlantic Ocean were within easy driving distance with beaches, fishing, and good seafood restaurants plentiful.

To top it all off, Gainesville had a good Class "D" professional baseball team. This, plus living practically next door to the Florida football stadium gave a fine sports dimension to my leisure time. If this sounds like the good life, it was.

I had been quite disappointed in not being assigned to Gulf Coast Military Academy but, all in all, I was having a most difficult time finding something to complain about.

Right away I was introduced to bass fishing, something new to me. A neighbor just down the street, an older man, on learning of my interest, offered to take me out to dinner to meet and get advice from two of the most noted bass fishermen in Florida. During the meal, they advised me as to the best reels as well as many little strategies and techniques used in this sport. They emphasized over and over the futility of buying all the lures in the sporting goods stores as most people are prone to do. They explained that most lures were made to please fishermen rather than the fish. They assured me that four lures that they named were absolutely all a fisherman needed and all the other lures simply served to clutter up your tackle box. If they wouldn't hit one of these, they said, then go home. I left that dinner with a firm resolve not to let myself get tricked into that expensive and fruitless habit.

Soon after this, Joe and I were on a lake casting away without the hint of a strike. After a while, we came upon another boat of fishermen. We asked them if they were having any luck. In the way of all successful fishermen, they allowed that it was a slow day but they had caught a few and proceeded to pull up a large string of pretty bass. Naturally we inquired as to what kind of lure they were using. The reply was that on that lake that day the only luck they had was with a black Wing Ding with a yellow tail. He emphasized it had to have a yellow tail. Joe and I could hardly wait until we were out of their sight before we cranked up and roared back to the fish camp to get us a black Wing Ding with a yellow tail, in fact, several. So much for firm resolve.

This same sequence occurred enough times so that by the time I left Florida I had enough lures to fill a Croker sack. Anyway, it is a fine sport and I enjoyed it greatly, including buying the lures. That was half the fun, at least. But I did finally accept the fact that it was the fisherman and not the lure that made the difference.

As bird season neared, Col. Edmonson eased into my office one day and suggested that he and I go together and get us a bird dog so we could slip away now and then to do a little hunting. I wrote Dad to see if he could get me one. He enlisted GoodHoney Rogers in finding one and we soon had a really outstanding dog to go to the fields with. Col. Edmonson enjoyed it at first but soon tired, so Joe and I made a lot of fine hunts together. Unfortunately, the dog was stolen by a deer hunter on the very first day of the next season.

Despite the disadvantage of the distance between Gainesville and Columbia and Cruger-Tchula, I kept up my courtship of Ree by mail, phone, and trips home. When I was home, not having a car, I would have to borrow one for a date. I can even remember that we went out in a pick-up one time.

One time we agreed to meet in Mobile for a weekend. Ree, Mur, and Pappy and I had a grand time together. I decided it was time to make my move. Ree and I went out to Bellingrath Gardens and I found the most beautiful spot I could and sat her down and proposed. The setting was on a short pier out in the lake. I felt the setting alone was worth ten points. I had assumed that she would give a "yes" or a "no" to a firm proposal. Instead I got a "Let me think about it". This threw me off my stride, but what can you do but wait? Anyway, before they left town, she answered in the affirmative. I believe she knew all along. Anyway, I left very happy and went back to Gainesville to buy an engagement ring.

I gave it to her on a date on my Christmas trip home. I can remember we were sitting in my new Chevrolet in front of Willie Henry's house in Tchula. We were double-dating with Henry Forrest and Sue. Henry Forrest had gone in to get Sue. (I think I lost ten points on this setting.)

I next invited Ree to come to visit in Gainesville to attend the Annual Military Ball. Ree and Mur stayed at Mrs. McCullar's house. Ree wore her engagement ring which served to announce the event to my community of friends. The orchestra was Claude Thornhill with his famous theme song, "Snowflakes". We agreed on a July 5, 1947 date for the wedding.

The AArmy was well aware that teachers had it easy in the summer months. so they had to think of something for us to do to earn our pay. They first summer they ordered me to Tennessee to give instruction to the State Guard. State Guards were formed by many states to take over the duties of the National Guard that had been federalized and then sent to war. They were composed of the old and the very young who were not eligible for the draft. I reported to the headquarters in Nashville, Tennessee and then was sent to the unit in Jackson, Tennessee. This trip was totally unnecessary. The war was over, and the National Guard units were returning to duty, so these people were just making time until disbanded. But like the sergeant said to the private, "Put on your raincoat and water the flowers, like the man said."

While I was conducting a training program, I met a most interesting person who started me on the road to investing. He was about the age of and reminded me of Mr. Buchanan in Cruger. He had been quite successful in life and his interest in me stemmed from my being from the Mississippi Delta. His wife was from the delta and he, being from the hills, was both amused and appalled by the ways of delta people. He said as often as not when visiting his wife's parents some friend would come by and say, "Come on with me. I need to get a cigar."

And, to his consternation, the man would drive forty miles and use up several gallons of gas to get a 10-cent cigar. This struck him as ridiculous. He had me out to dinner one night and gave me a most interesting account of how he had made a small fortune investing and trading. Then he fastened in on me...he wanted to know if I was saving my money. I very proudly told him I was and how much I had accumulated. Fortunately, I was able to save almost all of my salary during the war years. He then wanted to know what I was doing with it. I explained it was in a checking account in the Cruger Bank.

He almost fell off his chair...he was so upset. He gave me a long lecture on the benefits of earning interest and the beauty of compounding. He made it seem very close to a religious matter. The impression I got at the time was that letting money sit idly in a bank not drawing interest was very close, if not an actual sin, in most recognized religions, and certainly, it was a clear violation of the Constitution. He instructed me to make it my first order of business on my return to Gainesville to transfer that money from the bank to a Savings and Loan. I did, and to my amazement and delight, one year later I got a dividend check that was equal to one month's pay. I couldn't believe my good fortune, thirteen month's pay for twelve month's work. This taught me a lesson that has stood me in good stead and provided me with a most enjoyable and profitable hobby. [50]

In his Army service record, it is recorded that Earl Edwards attended two Army Schools during the year 1947. In all of the Army Registers from 1947 through 1962, Edwards is not recorded as having graduated from any Army School or course in the year 1947. Therefore these schools listed in his service record must indicate his posting to them as an instructor.

One is listed as TIS (The Infantry School—at Fort Benning, Georgia) and the course or major taken there is listed as Adv—Equiv (Advanced—Equivalent) with a duration of 36 weeks. In the following section of his memoirs, Edwards began the section with the heading of "THE FORT BENNING YEARS." However he gives no date of his assignment to Fort Benning nor of his duty while there.

The other school listed in Edwards' service record for the year 1947 is C&GSC (Command & General Staff College and the course or major taken there is listed as Basic—Equiv (Basic—Equivalent) with a duration of 16 weeks. Again this must indicate that Edwards was assigned to this school as an instructor. This may be the reference in the following section of his memoirs where he stated, "I was ordered to proceed to Ft. Riley, Kansas to serve as an instructor in a school for Army Officers newly assigned to ROTC duty." However, the Command & General Staff School should have been at Fort Leavenworth, Kansas, not at Fort Riley.

In his memoirs Edwards described his activities during the year 1947 thusly:

THE FORT BENNING YEARS

Early in the spring of 1947, I received a letter from Mother informing me that our family doctor, Dr. Gillespie, had diagnosed her with tuberculosis. Mother had been troubled for a year or so with a chronic cough and I knew she had been going to Dr. Gillespie for this. He had previously diagnosed the problem as just a chronic cough caused by nervousness or something. Apparently it had been getting worse and on one visit he told Mother that he had done everything except x-ray her chest. Why he hadn't done that earlier has always been a mystery to me since TB was very common at that time and a chronic cough was a normal symptom.

So, Mother said, "Well, let's get one." This x-ray showed TB. An earlier one would certainly have saved a lot of grief for our family.

Anyway, I took a few days leave and went home to see what we could do. Consulting with the doctor revealed that about the only treatment at that time for TB was bed rest and diet. So Mother was put to bed, and I got some help to cook and keep house so she could stay in bed all day. She still did not show any symptoms except her chronic cough.

I returned to Gainesville and soon received my orders for summer duty. I was ordered to proceed to Ft. Riley, Kansas to serve as an instructor in a school for Army Officers newly assigned to ROTC duty. I don't remember the dates, but I do remember I would be finished in time for our wedding scheduled for July 5.

While I was at Ft. Riley, I got word that Mother had begun to have fever and was not doing well. At first the doctor thought this was caused by something other than TB. However, on closer examination. it was decided that her TB had suddenly become more active and virulent.

When I got home I ran into real problems. I found that regular hospitals would not take TB patients because of a fear of infection, and I further discovered that the TB Sanatorium located in

Magee, Mississippi had an age limit of 50 for admittance for treatment there. Mother was 55.

Trying to think of what to do, I remembered that my old high school principal, Mr. Garland, was now a Representative in the state legislature. He had always been a good family friend (he would work on Saturdays in Dad's store) so I called on him for help. He told me to pick him up the next day (Saturday) and he would see what he could do. Ree and I picked him up and he directed us to go to the State Capitol. When we arrived, he and I went in and were soon in the Governor's office. Mr. Garland explained the dilemma to the Governor, and he got on the phone to the head of the Sanatorium.

At first he adamantly refused to admit Mother…with that, the Governor asked us to leave the room. I don't know what transpired then, but he soon called us back and said Mother would be admitted next day. Ree and I returned to Cruger and I arranged for an ambulance to take her there.

The timing of this caused us to move our wedding date from July 5 to the 7th. We had planned a very small wedding in Ree's home, so this change wasn't as big a problem as it could have been with a large church wedding. We did forget to tell the Glenn Walkers from Union, Mississippi so they arrived on the 5th, and we excused them from returning on the 7th. Ree and I were married on the 7th in a simple ceremony in her home. I will let her tell more about this. I had invited only Mrs. Griffin, Mrs. Estes, and Mrs. Flemming.

After the ceremony, Ree and I left for a honeymoon in Florida and Nassau in the Bahamas. On the way, we stopped in Magee to visit Mother. I remember her telling me that no doctor or nurse had been in to see her as yet. I thought this strange, but assumed they would shortly. I had assumed wrong. After we visited Mother in the sanatorium we left on our honeymoon. We spent the first night in Pensacola, the second in Daytona Beach, and then on to Miami. There we caught a plane for the Bahamas. In those days it was a quiet, peaceful little island run by the British. Except for the native population, you could think you were in England. We stayed in the Royal Victoria Hotel. It was old and old-fashioned in its ambiance. We had great fun going over to Paradise Island to swim. It had a perfectly beautiful beach.

After about a week, we flew back to Miami and proceeded on to Gainesville ... *It was at this point that I decided the honeymoon was over since the Pvt. began talking to me about a budget, how to balance a checkbook, the importance of entering dates, notes and names of people receiving checks, etc. I explained that while in Columbia my "friendly banker" and I had an agreement that whenever the money was getting low he was to call me on the telephone and tell me to stop writing checks. Since this system had worked like a charm for the two years while I was teaching in Columbia, I thought we could continue the practice in Gainesville. It took quite a bit of talking to convince me otherwise.*

I had rented a small house for us to live in. The morning after we arrived, I got up and sat around waiting for Ree to fix breakfast. She didn't seem to be making any moves in that direction, so I asked about the plans in this regard. She replied that she couldn't because her cook books had not arrived. Ree and I were slowly learning about each other.

Shortly we moved to another house and eventually lived in a third house, all rented, during our stay in Gainesville. Ree wanted to get a job teaching school and go to work right away. In my growing-up days, very few wives worked ... at paid jobs anyway. So I talked her out of it, which I now know was a mistake. I realize that she was bored being alone in a small house all day while I was away at work. After all, she didn't know anyone and there was very little to do in the small town of Gainesville in those days. Soon, however, she met some of the other wives and we settled into a pleasant routine of taking short sightseeing trips around Florida, going to ball games with Joe and Orlene. We often went out to watch the Class D baseball team play.

Fried chicken restaurants were just starting to be popular. There was one a short drive from Gainesville called Ruby's. We used to go there every Saturday night for supper.

Soon after we arrived, Ree expressed her desire for a puppy. We discussed the matter and agreed on a Rat Terrier. Each year we attended the Gator Bowl football game in Jacksonville. Ree found an ad in the Jacksonville paper for the sale of a litter of rat terriers. So, on our way to the game, we stopped by this place to look at them. When the man turned them out, one came directly to us and said plainly, "I'm it." We couldn't resist so we purchased him and took him home.

We eventually named him Nuisance because he definitely fit that title. I must say he was one of the most likeable dogs we ever had. He lived to about 12-13 years and traveled everywhere with us. Everybody we ever lived by became fond of him. He had a way of "conning" everybody into doing what he wanted. It was a sad day when he died of heart failure when we lived in Glen Carlyn in Arlington, Virginia.

Later in the summer, after we had returned from our honeymoon, I was reading an article in the Time magazine. It was about a new drug, streptomycin. That seemed to be a cure for tuberculosis. I remember it had a picture showing TB patients dancing in the aisles of a New York hospital celebrating their progress because of the drug. Because of this, Ree and I decided to return to Mississippi to see if Mother was getting this treatment. When I talked to the doctor who was head of the sanatorium, he told me in no uncertain terms that he did not believe in this new drug and did not propose to use it. Also, in talking to Mother she did not appear to be getting any sort of treatment. This, along with the fact that the doctor in talking to me emphasized the fact that I had to get used to the idea that Mother was going to die, left me puzzled and alarmed. Ree and I went on to Tchula and Cruger for a short visit.

Not knowing what to do, I went to Greenwood to talk to Dr. Gillespie. When I told him what I had found out in Magee, he gave me some startling information, He said that he had recently attended a medical meeting in Jackson and had run into the doctor who was head of the sanatorium. He asked how Mother was getting along. He quoted the doctor's reply as something like this..."the Governor forced me to take that old woman in against my rules and I will not treat her."

This was quite a shock to me, but it tied in with all I had seen and heard. I asked Dr. Gillespie if he had any suggestions. He replied that he didn't know anything about the streptomycin treatment, but he knew a young doctor who had recently left Greenwood to join a chest clinic in Memphis. He said, "I'll call him and see what he thinks." As it turned out, he talked to a Dr. Carr who was head of the Carr Chest Clinic in Memphis. Dr. Carr said he was so impressed with streptomycin that if I brought Mother to him he would give her a shot as she went down the hall to her room. So I quickly got an ambulance and headed for Memphis with Mother. As a matter of fact, one of the doctors in the Carr Clinic met us in the hall and followed us to our room and gave Mother a shot of streptomycin as soon as she was moved into her bed.

Mother remained in the hospital for some time until this new miracle drug brought her case under control. After she returned home, she had to continue the shots for some time. Mrs. Buchanan who lived about 100 yards away agreed to give Mother her daily shots...she had been a registered nurse before her marriage to Mr. Buck. This was a great help, and I was forever grateful to her.

To wind up a long story, the streptomycin brought Mother's case to a point where she could receive an operation as well as other treatments. The next summer (1948) she had a major operation in Memphis after which I arranged for her to be put on a train in Memphis and brought to Gainesville. I got Candy, our long-time Negro cook and family friend, to come with her. Candy returned to Cruger by bus, but Mother stayed with us until she was well enough to return home.

After that, she had other types of treatments and therapies until she was completely well. At one time I had to buy a car so she could be transported to Cleveland, Mississippi for a series of treatments. Also, twice, Ree had to return home to stay with her because Sonny and Fred had left to go to college, and we couldn't get adequate help. Altogether this took about five years, but Mother was eventually cured and became completely healthy again. She lived to a ripe old age of 84.

I have recounted this in some detail to show the kinds of problems you can run into in life and because of the impact it had on our early married years. It was a financial as well as an emotional drain on us at a time we would have liked to be free and easy. I can remember being able to charge about $6,000 per year off my income tax. That was a lot of money in those days and there was no Medicare or medical insurance of any kind. It all came out of our pocket. [51]

On July 15, 1948 Earl W. Edwards was promoted to the permanent rank of Captain in the Regular Army. Still on active duty, he continued to serve with the temporary rank of Lieutenant Colonel. In his memoirs, Edwards continued with the year 1948:

THE GERMANY YEARS

Early in the spring of 1948, I suddenly received orders to go to Korea. By this time all of our troops had been pulled out of Korea. I was to be the only American Officer in there and my duty was to act as a liaison between the U.S. and Republic of Korea forces. This was the summer that Mother was scheduled to be operated on. I called Dr. Carr in Memphis and informed him of this. He requested that I ask for a change or postponement of this assignment because he felt that I should be available during the operation in the event some decision had to be made. I also needed to be available to assist Mother for a while after the operation.

So I wrote a letter to the Pentagon (Infantry Assignment Section) and explained the situation. I got a quick call from my old buddy from the 11th Infantry. He quickly cancelled my orders and left me in Gainesville for another year. He also berated me for writing a formal letter and instructed me just to call him the next time I had a problem.

Early in 1949, I had another call from him to the effect that my ROTC duty would soon be up and that I was scheduled for an overseas assignment. He wanted to know where we would like to go. Ree and I talked it over and decided on Europe. Ree had no desire (then or now) to go to the Far East.

So, a little later we got our orders to Germany. We took a leave to visit home and soon departed for the New York Port of Embarkation. Nuisance couldn't go on our ship, so he had to stay with Mur and Pappy until he got a port call for a ship carrying animals.

He arrived in Heidelberg about a month later. We still laugh at the dejected look on his face as he was hauled into the station in a crate. He was understandably very disturbed by all that had happened to him since we had left. And was he a happy animal to see us again!

On our way to New York, I saw the first television set I ever saw. We stopped at a cafe to get a coke or something and they had one. The picture was so poor that I was not very impressed.

Later, when we stopped to visit Ree's cousin in Washington, we were able to look at a much better set and the picture was good. I got to watch the U.S. Open Tennis matches for the first time. I was impressed.

When we arrived in New York, we received a change of instructions. We had to catch a train and go to Boston, Massachusetts to board our troop ship.

While standing in line to get aboard the ship, we met Macon and Miriam Hipp and Ernie and May Janes who were going to be our shipmates. On the voyage over, we became good friends. Since all of us were eventually assigned to Heidelberg, we kept up our friendship and have until this day. The trip over was pleasant but uneventful. Just lazy days, eating and sleeping our way across the Atlantic.

After we landed in Bremerhaven, we took a train with Pullman-type rooms and made the all night trip to Bad Kissingen.

It was very exciting to be going down through such a strange (to us) foreign country. I don't think we slept a sink. We kept looking out the window at the countryside and listening to the strange language of the people when we stopped in the towns.

Bad Kissingen was set up as temporary quarters while we waited for our orders. It was a beautiful resort town and we enjoyed walking about town and looking in the shops. Somebody advised me NOT to buy anything there since it was the most expensive place in Germany. I cautioned Ree about this and, finding it worked, I tried to use that device every place we went. It didn't work very long, to say the least.

When my orders arrived, I found I was assigned as a Battalion Commander in the 1st Infantry Division. I knew this was a first class assignment since the 1st Infantry Division was the premier division in the US Army, and especially in the US Army for Germany at that time. Nevertheless, I was disappointed. I knew this meant many, if not most, days would be spent in the field. Also, I knew it would be difficult to get off to make trips around Europe.

So I headed out to Heidelberg to see if I could get some other type of assignment. I knew that Autrey Maroun and Willys Pearson, old 22nd Infantry buddies of the very early days, were stationed there. We got together for lunch, and I told them what I was trying to do. They both had key assignments in EUCOM headquarters in Heidelberg, so they immediately suggested I leave the matter in their hands.

They also explained the problem would be to get Lt. Gen. Huebner to approve my release from the 1st Infantry Division. He was the wartime 1st Infantry Division Commander, and at this time, he was the Deputy Commander of EUCOM Headquarters. They told me to return to Bad Kissingen and await developments. Sure enough, in a short time, my orders to the 1st Infantry Division were cancelled and soon I received orders to report to the Intelligence Division of EUCOM Headquarters.

Ree and I were assigned to live in what was now called the Truman Hotel, an old bombed out German Hotel located a short way down the Neckar River in Mannheim. We used to say we were "rooming at the Truman."

I started to work in the Intelligence Division and began to slowly learn the ways of a large headquarters. EUCOM Headquarters was a joint command for all of the US Army, Navy, and Air Force units in Europe. This was quite a jump for me…from a Regimental Headquarters to this high level of staff work.

Several months later, just as I was beginning to get my feet on the ground, I was reassigned. The two most prestigious assignments in Germany for Lt. Col. were in what was called the Command Building. This housed the Commander-in-Chief (General Thomas Handy) with a staff of several generals (Huebner, Noce, Johnson, Williams) and two Lt. Cols. The first floor housed a secretariat that did all the paperwork. Willys Pearson worked in this section. The Lt. Col. Spot that served the personnel, financial, civil affairs, and supply part of the business came open. Willys and Autrey went to work without my knowledge to get this job for me. The first thing I knew, I was called over for an interview and wound up in the job. I remained in this job for the remainder of my tour in Germany. This proved to be one of the most difficult jobs in my entire career. It demanded long working hours and a high pressure environment at all times. It was way above my previous experience, and I suffered some rough times learning how to cope with the flow of work and pressure.

The first nice bonus was that I was required to make a tour of all the major military installations in Germany. The purpose was to meet the commanders and staff of each installation, be briefed on their mission and problems, and get familiar with the physical layout so I could understand what they were talking about when they called our headquarters. Ree went with me and, except for the fact that she was pregnant and suffering from morning sickness, we had a grand tour of Germany.

Ree and I thoroughly enjoyed our three years in Germany. We

took frequent Sunday rides, and we also took full advantage of our leave time to make trips. I'll let Ree tell about those.

I also enjoyed being again with Autrey Maroun and Willys Pearson. Autrey was not married, but Willys had a wife named Nellie Bell. We went bowling together and one time we made a weekend trip to Berchtesgaden together. Ree and I have never forgotten the overnight ride on a train through the Alps from Heidelberg to Berchtesgaden. I still remember us having breakfast in the dining car as we rode along in the Alps. The views would be breathtaking. We also enjoyed a continued association with the Janes, Hipps, Burketts, and Glenns.

After living in the Truman for a while, we were assigned to some newly built apartments very near the headquarters. These apartments were very nice and well furnished. It was while we lived here that Lyn was born.

Ree decided we needed to go to the hospital in the early hours one morning on May 29, 1950. Lyn was born a few hours later, premature by six to eight weeks. I was scared to death because I had no idea she could live, having been born that early. But Lyn was tough, and she meant to make it. We were unable to bring her home for a while because of her early birth. When we did, the doctor had hard, specific directions about her care. I remember with great clarity his admonition that if she didn't get the prescribed number of ounces of milk each day, she would die.

I sometimes (he had written often but Ree made him cross it out!) had the 2 A.M. feeding and I can remember Lyn getting sleepy and how frantic I would be trying to keep her awake and feeding. Sometimes the nipple would stop up ... what a trying time!!!!!!!!

But Lyn did well and gained and grew each day. I can also remember with great pleasure when she got strong enough to raise her head above the rail of her bed and grin at me when I came home from work.

The Army had a policy of assigning a German woman to each house, supposedly to help maintain the house but actually we used them as housemaids. Ours was named Else Osterhage. We called her "Ato". She was a wonderful person. She was about 50 years of age; her husband had been killed in the war. She helped us in every way, and she loved Lyn so much that she tried to do all of the taking care of her, much to Ree's dismay. She was able to take care of Bruce some, but we left Germany soon after he was born. (Ed., Bruce Alan Edwards was born on July 20, 1952.) Along toward the end of our tour, she began to feel bad. I can remember her coming to me and asking permission to take a drink of whiskey to relieve her pain. Sometime after we returned to the U.S. we got a death notice. Apparently she had cancer.

After living in the apartment about a year, we had a chance to move to a lovely house on Ludolph Krehl Strasse. It was across the Neckar River on the side of a mountain. We really enjoyed living there.

Charlie and Helen Berkowitz, who had lived in the apartment just across the hall from us, also moved to the house just across the street from us. Next door we had Fritz and Anita Fredericks. They had a little girl the same age as Lyn called Dodo. She and Lyn became good buddies. We used to have our Easter Egg hunts in their yard and we all had fun watching them stumble around looking for eggs.

From time to time, almost all the officers living on this street would go into the field for some kind of training maneuvers. The Germans knew this and would use this time, with husbands being away, to steal cigarettes, whiskey, and coffee. This is all they wanted. Since all the wives knew I was home and could get to their house before the MP's could, they got to calling me to come when they detected a break-in attempt. I got to where I kept my gun and a flashlight by my chair so I could respond quickly. I also always took Nuisance with me on a leash. When I got to the yard, I would turn him loose and he would ferret out the thief, usually hiding in some bushes when he heard me approach. Most would simply run away, but I did actually arrest three of them.

In January or February of 1951, Ree took Lyn back home to visit her grandparents. May Janes made the trip with her on her way for a visit to her home in Kentucky.

Then, in that summer we were able to talk Mur into coming over to visit us. Unfortunately, she took a bad cold on the way over and it took her some time to recover. But, when she did, we took a trip to Italy. We visited Florence, Rome, and Venice and had a grand time. We visited all the usual tourist sights. When we were in the larger cities, we stayed in large hotels in the center of town but the rest of the time we stayed in small family run hotels called "pensions". Italian people love babies, so each time we stayed in the small family hotels the wife of the owner would come and get Lyn and take her into their family quarters and play with her. Lyn ate this up.

Mur didn't cotton to high places, so we had a time getting her over the Alps and back. Mur didn't care how high her airplane flew, but when she was in a car, high places simply scared her to death. When we were in Venice we purchased one of the beautiful table cloths that we still have.

The next summer (1952), Mur came over again. This time we took several trips, one into Holland at the best time of the year to view the tulip crop. I never will forget the first time we came upon a 40 or 50 acre field of red tulips in full bloom. It was a sight to behold. We stayed in a small hotel on the coast and toured around the country from there. We visited the town of Edam, which is famous for the production of cheese. I stopped at a store that sold nothing but cheeses. What an experience for a cheese lover!! Our car barely made it out of town before we stopped for a feast of cheese and crackers.

We also visited Keukenhof, which is a large estate used as a showcase for all the tulip bulb producers in Holland. Each company is allowed a small plot to showcase and sell their bulbs. It was in full flower, so Mur went as berserk over the bulbs as I had done in the cheese market. She selected many varieties to ship home to Warfield.

The next trip we took was to France, England, and Scotland. First, we visited Paris for several days. We toured the city, enjoyed the fine restaurants, and went to see the show at the famous Follies Bergere.

After Paris, we made our way to England. We crossed the channel on a ferry that landed at the white cliffs of Dover. We spent the first night in a lively little hotel right on the white cliffs. After that we headed for London. We had reservations in a fine hotel right in the heart of town. We toured the city, visiting all the usual tourist sights. On Lyn's birthday, the 29th of May, we went strolling in St. James Park. I was able to get tickets to the famous Palladium for a concert by Jimmy Durante. Also one night we went to the Drury Lane Theater to see the American production of *South Pacific* with Mary Martin. In the old-fashioned British way, they served tea at intermission time. We all enjoyed this very much.

One day we visited the famous silver vaults as they were called. As I remember, these were stores one to three floors below ground displaying and selling silver offered by estate sales. All of it was antique or, at least, very old. Every one of our friends who visited London strongly urged us not to miss the chance to buy some fine

English silverware. Most everyone bought a full tea service consisting of a large tray and several tea, and cream pitchers. We found the price, while most reasonable, was just too much for us...most costing several hundred dollars. So we settled for a large tray...the one we still have...that cost $40.00. With the top, it cost $50.00 but they agreed to take the top off and let us have the tray for $40.00. We were that close on money.

After London, we headed for Scotland. We had planned to go well into Scotland, but, on realizing that would be into the Highlands, we decided Mur would be very unhappy. So at Edinburgh we turned back. On our return trip we visited Belgium to see what it looked like.

THE FORT LEAVENWORTH YEARS

We departed Rhein-Main airport at Frankfurt, Germany on the 23 September, 1952 for our return to the USA. Since our household goods had been packed up and would be shipped direct to our new station, we wouldn't get to see or use any of it until the end of our leave at home and our arrival there. For this reason, Ree kept out all the items that we would need in the meantime that we thought we could carry on our persons. We looked like refugees from a war zone with bags hanging all over us. When we departed the lounge to board the plane, Lyn was so loaded down she just couldn't navigate, and we were too loaded down to help. Luckily some kind person, seeing our plight, pitched in, and got Lyn on the plane.

When we arrived at our plane, I was appalled. It was an old, dilapidated four-engine job that belonged to some fly-by-night private carrier that was under contract to move military personnel about for the AArmy. The outer skin had patches all over it, so I was uneasy at the outset at the prospect of crossing the Atlantic Ocean in this rig. We landed in Shannon, Ireland and then headed for Newfoundland, Nova Scotia, landing briefly there before heading on to New York. When the pilot announced that we were at the "point of no return" over the Atlantic, I pointed out to Ree how oil was oozing out of the engines and streaming back across the fuselage and expressed my strong desire to return anyway. Ree didn't approve of that type of gallows humor and expressed her strong desire never to ride anywhere on an airplane with me again.

We arrived in New York on the 24th of September. As soon as we could complete the paper work and get possession of our trusty Oldsmobile, we headed for home. It was a three-day trip over two-lane roads (no Interstate then) with no air-conditioning. It was a long, hot ride with two small kids. Along the way, Ree had to be up late each night after we got Lyn and Bruce to bed, making formula for the next day using the most rudimentary facilities. (Ed Note: Bruce was born to Earl and Ree on July 20, 1952.)

On the last day out, a Sunday, we stopped early on at a restaurant to eat breakfast and ice up the formula chest. Later on in the day as we neared Columbus, Bruce got hungry and made it known in the usual fashion that he was ready for a bottle. A quick search of the car revealed that we had left the ice chest full of formula back (a long way back) at the restaurant. With Bruce venting his anger and frustration at our inept management, we rode around Northport (just outside of Tuscaloosa) looking for an open drugstore so we could purchase all the equipment necessary for making and storing formula. Then circling on to find a restaurant that would allow Ree to get in the kitchen and restock our supply. Bruce was simply not impressed with our way of doing business. We finally got the job done and headed home. These are the kinds of things that happen in life that you try to forget but this is one burned in our memory.

We finally made it home for a warm welcome and a nice leave and visit with our families. We were not due in Ft. Leavenworth until the 17 October 1952. Nuisance had been shipped on a pet ship and arrived in Tchula soon after we did.

When we arrived at the famous old military post, we were assigned to Quarters 219 Pope Avenue. It was a large, red brick house of a standard design for the era it was built in. It was roomy and very comfortable, only a short walk to work, so it was most convenient. I could even walk home to lunch. We quickly learned that a number of our friends from past assignments were stationed there. This is one of the more pleasant aspects of military life … at each new post, you rejoined many old friends and made some new ones. Glenn and Margaret Walker were there and were especially helpful in giving us a hand as we settled in. [52]

Earl W. Edwards was promoted to the permanent rank of Major in the Regular Army on October 31, 1952. He continued to serve with his temporary rank of Lieutenant Colonel and was assigned to duty as part of the staff and faculty of the Command and General Staff College at Fort Leavenworth, Kansas. His Military Occupational Specialty was changed to 2728 (Military College Faculty Member) on July 6, 1953. His unit at Leavenworth was officially known as the 5025th Army Service Unit, Command & General Staff College. Edwards continued in his memoirs of the days at Fort Leavenworth:

Ft. Leavenworth is the home of the Army Command and General Staff College, to which I was assigned as an instructor in Department V. It was also the home of a number of prisons, including the famous Federal Prison. At home, people thought that when you were sent to Leavenworth you had been sent to prison.

I was assigned to a small section of three officers including Lem Downs and Gus Gustaffson. We covered Personnel and Supply subjects. All three of us later on wound up in the Personnel section of the Pentagon and Lem Downs was assigned as one of my assistants, much to my delight. He is now retired and lives in Tavares, Florida near his old home. We recently visited them there.

To my dismay, I was assigned to write and teach two courses that were taught as a two-hour lecture to the entire student body of about 500 students in a large auditorium. This experience was not something I wanted or enjoyed, but I am sure it helped me when, at a later time, I had to make so many speeches during my tour as Executive Secretary of the MPSA.

As I have mentioned earlier, we saw a little TV in its early stages on our way to Germany. By now, three years later, it was in full swing, and it had everyone enthralled.

Every house had a TV set, and it was in absolute control of all social life. As we started out to call on and pay our respects to our various friends, we found them all so completely preoccupied with TV that a visit usually turned out to be a joint TV watching. They were simply not willing or able to give up their viewing of a favorite program to accommodate a visit by us. Since we were not into these programs, mostly sit-coms, we were simply appalled at their lack of courtesy and interest in our visit. This kind of turned us against TV. We decided if it had that effect on social life, we'd just as soon forego the pleasure. This lasted until two events occurred that made us take a second look at TV.

The AArmy had decided to take on a famous, or more correctly infamous, senator named Joe McCarthy. He had been attacking the Army (and others) as being soft on Communism. He had even inferred that General Eisenhower was a closet communist. The AArmy had hired a famous attorney, Joseph Welch, to take him on in some Congressional hearings he had called. Since this was of great interest to me and scheduled for television, I felt I had to see it. Also, at about the same time, the baseball World Series was scheduled for TV showing.

This did it…Ree and I gave in and bought our first TV set, but we did set some rules. We put the set in an unused bedroom upstairs with two chairs. We were determined that it would not reside in our living room where it would dominate our family life. So when we wanted to see a particular program, we would march upstairs, watch the program, and return to the living room when it ended. This way TV never dominated our lives in the way it has come to do for most people. We still notice that when we stop to visit somebody, they rarely cut the TV off. The more polite will at least cut the sound down so we can converse, but their eyes can't help to stray in that direction occasionally. It leaves you wondering just how welcome or timely your visit was. To this day, we have never put a TV in the living room…and that is the reason.

We were in Leavenworth for two years and we enjoyed all of it. One of the more pleasant things to do occurred only once a month…PAY DAY! We would load up and go over to St. Joseph, Missouri to eat at a famous cafe located out in the stockyards. It was called, aptly enough, The Hoof and the Horn. Ree and I would eat a T-Bone steak, Lyn a hamburger and Bruce a bottle of milk. Those steaks were out of this world and so large they lopped over the sides of the plate. With an ample portion of French fries, a good salad, and a bottle of draft beer (for me), it was an event to savor and remember. We would complain all the way home at being too full…and a five-dollar bill covered the check and a tip. We often recall those nights out with pleasure.

Our culture was enhanced by going over to Kansas City to take in a program at the Starlight Theater. Programs were performed in an open air amphitheater under the stars. Our most memorable occasion was attending a performance of Sigmund Romburg's "*The Student Prince*". The singers were superb and the songs a delight. One of the songs "*Golden Days*" is still one of my favorites...all of this on a perfect evening full of stars and in the company of good friends.

We often had guests in for a meal. Some of the students we had known in previous assignments so we would have them over usually for breakfast on Sunday mornings. Jack Eakin was one we especially enjoyed. He came often. Jack had been a policeman in his pre-AArmy days in Birmingham, Alabama and he had an endless number of hilarious stories about those days. Our courtesy here paid off...later as we'll explain. Macon Hipp, one of our old Heidelberg friends, came by as well as Fred Kent from my service in the 22nd Infantry Regiment.

I soon became involved with a group interested in bird-hunting. Glenn Walker, Jake Powell (from the 8th Infantry Regiment, 4th Infantry Division), Bobby Cameron, and I formed a small bird-hunting fraternity. Jake had been given an outstanding bird-dog on his return from WWII by his father. He bred this dog to a World Champion and sold each of us a pup from this litter. This particular pup proved to be worthless and was no end of trouble. I didn't get him until he was a few months old and when I brought him home Ree and I needed to go somewhere so I left him shut up in the kitchen. Unfortunately, I also left a brand new radio that I had just purchased as a present for Ree on the kitchen table. When we returned, he had pulled the radio off the table and made a shambles of it. Not an auspicious beginning...and it didn't get much better with time, so he stopped being my bird-dog pretty soon.

Later on, Bob Cameron purchased a puppy from a famous kennel. I remember he paid $600.00 for an 8-week-old puppy...an unheard of price at that time. This pup, as he grew up, became very difficult to control. When we would take him out in the field for a workout, he would simply run away, and it would often take days to locate him. One day Bob told me he had an idea of how to establish control of such a dog and break him of this abominable habit. He said what if he put a collar on the dog that had a built-in electric shock capability, and he had in his hand a device that could activate the shock from the collar at his pleasure. Could he not use it to teach the dog that he could not escape by simply running away? I allowed as how it might work, so one day Bob informed me that he had looked into the feasibility of this idea and had found an electrical engineering firm that would design and build such a device for $24,000.00.

He expressed the belief that such a training device would be a popular thing for owners of hard-to-control dogs and that there would be a large market for it. He asked if I would like to put up $12,000.00 and go into business with him as a partner. Unfortunately, I had to inform him that available capital would not permit me to join him in this venture. Anyway, Bob went it alone and became the inventor of the first of this now very commonplace and useful training device. In later years, Bob told me he made a lot of money out of it but eventually had to sell out. The reason was that people kept making "copy-cat" models which kept him busy in court suing them for patent infringement. He said he simply couldn't afford the time and attention to this business and still remain in the AArmy. Eventually Bob did retire at about the same time I did and became a stockbroker. Ree and I visited Betty and Bob in recent years on one of our trips out West.

Eventually I acquired another bird dog, so with two dogs and a cat, we realized that on some occasions we would have to transport them … either to the vet or on trips. So I looked around and found a small second-hand trailer that we could hook behind the car. Unfortunately, it was only a frame, which meant that I had to build a little cage for it to carry them. I reasoned that this would really be no great problem but the more I looked into it the more I realized that this project was just a tad above my limited carpenter skills. But it just happened that our next door neighbor had renovated and rebuilt a large house trailer in the space between our two houses. Ree had noted that he was obviously a high class carpenter, so she suggested that I would be well advised to seek his advice.

His name was Arnold Taynton and it turned out that he had commanded the 70th Tank Battalion which often supported my regiment in combat. So I went over and expressed my great admiration for the work he did in reconstructing his house trailer. I explained to him what I was trying to do and asked only that he help me develop a list of the lumber and tools I needed to do the job. He quickly and readily composed such a list so off to the lumber company I went. On my return I placed the bare-bones trailer right between our houses along with the materials needed.

The next day I put on my work clothes and after calling him on the phone to get some advice on just how to get started (What is the very first thing I must do? Then what is the second thing I must do? etc.) He gave me very specific and detailed instructions for which I was very grateful. So then I went out to get started. It is well known in the family that my carpentry skills are very limited, so it should be no surprise that as I tried to carry out Arnold's complicated instructions that I looked very awkward as I picked up first one plank and tool, then another. Of course Arnold was watching out the window, so he began coaching from the sidelines but to little avail, even though I was doing my dead level best to follow his instructions.

After a few minutes of this, he came boiling out of his house, abruptly pushed me aside, and proceeded to build me a first class trailer cage. It is very difficult for a professional carpenter to watch an amateur butcher perfectly good lumber and abuse tools.

This experience taught me a valuable lesson for use in later life in getting expert help on a project in areas that I was deficient in. Ree said then and has mentioned ever since, that it was my plan from the very beginning to con him into building it. She has never stopped shaming me about it. Nevertheless, we got a first class trailer built, and Arnold was very proud of the job he did. So there was benefit to both sides … and besides, I thanked him profusely.

One day we suddenly remembered that Leo Lemmerman from the Florida days should have completed his studies and returned to Kansas by now. So we checked and found his telephone number and called to see how he was getting along. Instead of Leo, his wife, Naomi, answered. I had only met her very briefly one time on a visit she made to Gainesville, so it took some explaining for her to understand who I was. When she did, she simply dissolved into tears and slowly began to explain to me that Leo had never returned. Instead, he had apparently fallen in love with some girl in Gainesville and divorced her. This had simply wrecked her life. It seems they had grown up very poor, had married at about 17 years of age, and she had spent most of her life working trying to get him a good education so they could have a good life together. Now, all of that was blown away.

As she sobbingly explained all of this to us, we didn't know what to say or do. Finally, we suggested that we come and visit her on Sunday. We did and she seemed to appreciate it so much that we invited her to visit us in Ft. Leavenworth. She did several times and seemed to thoroughly enjoy playing with Lyn and Bruce. We kept up with her for years; one time she visited us in Washington.

She was a very nice person, and we couldn't help but feel she had been assigned a tough row to hoe.

One time Mur and Pappy came for a visit, and we thoroughly enjoyed showing them the country, especially the Kansas City stock yards. We also went on trips home each Christmas and summers.

Army pay during these years was never generous; it was just about what you needed to live on. This, plus the fact that we had to help Mother and Dad with a monthly check, and an occasional medical expense, left us always kinda' close to the edge. The AArmy had a system of giving small cost-of-living increases called "Fogies" for some reason about every two years. It usually amounted to 18 to 20 dollars so we would just let this go into our checking account and it was so little we never really noticed any difference. So I hit on the idea of using "Fogies" as a saving rather than let them be absorbed in our living. My way of doing this was to buy more insurance which I felt was inadequate at the time. Instead of increasing it by the exact amount of the "Fogie," I would up it to 25 dollars or $27.50, etc. As time went by, instead of our net pay going up, it was gradually going down. One day Ree caught on to this slight scam and called my hand. She said, "One more Fogie and we can't pay the grocery bill!" We asked our insurance agent to come in and give us a review of our situation. His name, I remember, was Madison M. Letts. He concluded that we were insurance poor and could well rest for a while. Anyway, this insurance came in good stead when we wanted to buy our present house. The cash value of all our policies was such that we could pay off Mrs. Winstead and assume the original mortgage, which had a very low 4 ½ % interest rate. Our payment would then be less that it would have been to rent a house.

During our stay in Leavenworth, Lyn began to develop a

tendency toward respiratory problems. We never knew the exact cause. She would get an attack every now and then on a regular basis and be sick for several days. The pattern was for her to worsen to a point that she would vomit. Then she would be relieved for a while until she had to do it again. If it happened at night, we couldn't sleep as we had to sit up with her until she vomited, get relief, and then go back to sleep…and so could we. Sometimes this would take an hour or so.

On one of our many trips to the doctor, he suggested that he could give us some ipecac to give her, which would speed matters along. Lyn called vomiting "bumbling". She would say, "I got to bumble." We learned to give her a dose of ipecac and shortly Lyn would say "I think I have to bumble." We would laugh and say, "We think so too." So we'd quickly head for bed and a few hours of sleep.

As Bruce grew up, we began to notice a vast difference in their personalities. Lyn had always been without fear or any idea of using caution in whatever she did. She would attempt to climb up on anything, jump off or into anyplace. On trips if we were staying in a motel and she got near the pool, she would just jump off in the deep end and promptly go under. We would hurriedly reach down and pull her out and, if not watching, she would do it again. We had to learn to watch her like a hawk. On several occasions when I was negligent, we had some near disasters. Ree, at one time, especially aboard ship, would put a little harness on her and hang on. In Heidelberg we lived in a two-story house so we had to put a gate at the top and bottom of the stairs because Lyn, who couldn't even walk very well, would attempt to walk down the stairs and just tumble all the way down, scaring the wits out of us.

So when Bruce came along and began to crawl about at Leavenworth, we promptly got out the gates in anticipation of the same kind of behavior from him. We quickly realized that they were a different breed of cat. When Bruce started walking, he would get down on all fours about ten feet back from the top of the stairs and cautiously crawl backwards to and down the stairs. No problem here. This difference in behavior played itself on out the same way as they grew up. One bold and reckless, the other cautious and careful. How can a brother and sister be so different?

While in Leavenworth we made the first credit purchase we ever made (aside from a car)—a piano. We paid it off over a period of several years. Ree had taken piano lessons in her youth, and she wanted Lyn to have her chance. I guess every parent hopes to discover a talent for something…music, art, whatever…in a child, and they want to take whatever steps are needed to make sure any such talent is not overlooked. Alas, as is so often the case, this was not her talent. We gave her lessons for some time and finally came to the inevitable conclusion and gave it up. In music or art, you either have it or you don't. Lyn didn't.

I learned one important lesson in car ownership while we were here. We still had our Oldsmobile that we had purchased in the Florida years. It now had more than just a few miles and was beginning to give hints that it was at or near trade-in time. The motor began to run a little rough, so I thought I should get a tune-up, to include spark plugs before taking it in for a trade. I felt like a smooth running motor would help us get a good trade-in value. So we made a trip to Kansas City and visited the Greenlease-O'Neill Oldsmobile dealership. In the repair shop I suggested a tune-up and a change of spark plugs and then, as an afterthought as we were leaving to go shopping I said, "Oh, yes, and if you see anything else that needs fixing, go ahead and do it while you are at it." I now know that is all a good mechanic needs to give you an overhaul. When we got back to pick up the car the bill was beyond my imagination. He had found five or six items that needed repair or replacement. The bill put Ree and I in a state of shock. We decided that now we couldn't afford to trade…we'd have to keep it a few more years to work off the repair bill. I will say, it ran like a top until we did trade.

One of our cost-cutting measures that I hit upon at this time was an idea to cut Bruce's hair. He hated going to the barber shop with a passion and haircuts were expensive, so having seen some haircut sets for sale in various magazines I reasoned that this was an excellent way to save money and time. The set came in and shortly after, much to Bruce's dismay and with great reluctance, I got him to sit up in a chair and get ready for his first home haircut. I soon realized it wasn't quite as easy as the pamphlet depicted, and I was getting a little disgusted because it wasn't looking too good. I remember having to caution Bruce several times to sit still and not move his head about so much.

Anyway, whether because he moved suddenly or because my barbering skills were somewhat akin to my carpentry skills, I suddenly snipped a small (very small) piece of his ear off. Bruce instantly launched into an Oscar-winning performance to indicate pain and suffering. I never have believed that the extent of these histrionics were in any way justified by the slight damage done. I have always believed that his primary purpose was to embarrass and humiliate me.

Anyway, I took him over to the outpatient clinic at the hospital for repairs. The doctor gave me one of those "I don't believe this!" looks when I had to explain how it happened. I had to tell the truth because Bruce was sitting there and the look on his face told me he wasn't in the mood to let me off the hook. The doctor even implied that the AArmy might not be willing to pay for it. That was the one and only time I was willing to put myself out to cut a kid's hair to try to save a little money. Thereafter I accepted my limits and paid my dues.

Normally I would have been stationed at Leavenworth for three or four years if I behaved myself. But at the end of the second year, I received orders to attend the Army War College in Carlisle, PA. So, like always in the AArmy, after a while it was time to move on once again. [53]

(While at Fort Leavenworth, Edwards attended the Command and General Staff Regular Course, graduating from that course in 1954. In his service record, that course was recorded as lasting for ten months. On his final separation document the same course was recorded as lasting for forty weeks).

On January 31, 1955, Earl W. Edwards was promoted to the temporary rank of Colonel (AUS).

Edwards' next assignment, to the Army War College in 1955, was also recorded in his service record as lasting for ten months and in his final separation document as lasting for forty weeks. In his memoirs, Edwards described his War College days:

THE WAR COLLEGE DAYS

In the peacetime Army, you simply train for war. As an integral part of the process, the Army has a whole series of schools, some for officers, some for enlisted men, some for command and staff, and others for specialized purposes. You start at the bottom and at the end of the line is the Army War College. Here, you try to put it all together, strategy, tactics, how the industrial base fits in the scheme of things, etc. It often seemed that all you did was go to school.

The Army War College, the pinnacle of the system, was located at a post called Carlisle Barracks in Carlisle, Pennsylvania. This historic old Army post was established during the Indian War days. Here, also, had been the home of an Indian college whose team was called the Carlisle Indians. Jim Thorpe, one of the most famous athletes ever, played for this team before he turned pro.

Attending the school was a most pleasant experience. There were no real exams to take, but you did have to do studies on assigned subjects and present papers to be graded and approved by the faculty. The daily schedule started off each day with an eight o'clock lecture by some well-known speaker. After that we were off to a schedule of seminars and study-groups on a wide variety of subjects. Often people from government and industry were brought in to add their knowledge and expertise to our program. One time I was assigned to work with Senator Strom Thurmond for several days on a political project.

We also took field trips from time to time to visit various industries that form a base of support for the war effort. One very interesting one was to visit the Bethlehem Steel Company at its famous Sparrows Point plant near Baltimore, Maryland.

The area in and around Carlisle is very historic with many interesting things to see and do, so our off-time was most delightful.

By the time we made the trip from home to Carlisle, the dogs and a cat had thoroughly adapted to travel by trailer. They all rode together and must have made a funny sight as we would see people laughing as they passed us by. To give them some relief on a long trip, we would try to stop every two hours or so. I would find a side road that bordered an open field or pasture and pull in. When I opened the cage door, both dogs would leap out and race madly around the field until I blew the whistle signaling for their return. I usually gave them ten minutes and they always obeyed my signal promptly. They would load up and off we would go again. Most amazing, when we spent the night at a motel we would let the cat out and leave him out overnight. I was secretly hoping that he would fail to return. He was a big, male cat and, as I saw it, nothing but trouble. But not once did he fail to show up when we began to load up in the morning.

One time when the motel was located near the edge of a town, I watched him slowly make his way into town just at dark. I just knew this was the last I would see of him, but magically he appeared out of the bushes when I cranked up the next day.

The last day out, for some reason, we let the cat ride with us in the car. He was lying between Ree and me on the front seat, sound asleep, when a big trailer truck passing by made a loud backfire. This caused him to take a flying leap to get out the open window. Somehow, in a reflex action, I managed to catch him in my hand in mid-flight. So, when we pulled into Carlisle I had to head for the Outpatient Clinic to be treated for some long, deep cat scratches. I really loved that cat.

For living quarters we were assigned to the Stanwix apartments. These were probably of wartime vintage and built by the very lowest bidder (and I am sure he cheated!) They were so poorly constructed that, when the wind blew, (which was most of the time), it blew right through the apartment. We nearly froze to death most of the time. One time when we were especially uncomfortable, Ree protested to the building superintendent and he explained there was nothing he could do because the wind was blowing. However, this all being true, Ree always said we actually had fewer health problems than we did in our very warm quarters in Leavenworth.

It was while stationed here that I was promoted to full Colonel on 31 January 1955. A promotion is always a happy event. Not only do you feel you are making progress, but your pay goes up. As Ree often tells the kids when we pull rank and take the best of whatever "Rank has its privileges " …and the more rank, the more privileges.

Lyn was getting big enough that we decided to let her catch a bus by herself each Sunday morning to go to Sunday School. Out of her sight we would get in the car and follow her to make sure she got off at the right church and back on the bus when it was time to return. She really thought this meant that she was grown up at last. After a few trips, we felt confident enough not to follow her.

Bruce developed a good, working friendship with a boy named Buck who lived in an apartment nearby. Each loved to play with the other's toys. As often as not, when they received permission to visit with each other, they would each start out for the other place. When they met about half-way they would stop and talk a little and then each would proceed to the other's apartment and happily play alone with his toys. We thought this procedure was hilarious.

It was here that I made my one and only pheasant hunt and killed my one and only pheasant. I say I killed it but, really, my hunting companion and I both shot at it at the same time. Since I was the nearest, he insisted it was my kill. I happily accepted his reasoning. To Ree's dismay, I brought it home for cooking. It was here that I began to realize that Ree had no objections to me hunting or fishing, so long as I didn't bring anything home.

During our stay in Carlisle, John and Doodle were living in Ithaca, New York while John was attending Cornell University getting a doctor's degree in Entomology. They were living in an apartment off campus, so we visited them one week-end. Johnny was just a small boy. They served pizza for supper …my first taste. Later when John got his degree, they stopped by to visit us on their way back to Florida.

Bob and Betty Cameron were now living at West Point, New York. Bob was an instructor at the U.S. Military Academy. We took this opportunity to visit this famous institution. It is an impressive sight, located as it is on the banks of the Hudson River. We were impressed.

Pennsylvania, at that time, was home to quite a few small family-owned furniture factories. Usually, it was just a father and son operation, and they often owned the woods where the lumber came from. Typically, it was black walnut. So, on learning about this, we prowled around until we found one we liked and placed an order for two end tables. It took a while for them to be produced, but they were well worth the wait. They were beautiful. However, for some reason, we sold them when we left Lebanon.

There was the usual Farmer's Market and we made regular use of it to stock our kitchen. It was quite a treat to visit and view all the wonderful eatables this area produced.

I made friends with a fellow student named Bob Evans. I think the fact that we both smoked cigars was what initially attracted us to each other. It turned out that he was an heir to either the McCormick or the Deering fortune, with all that this implies. Anyway, he was a real cigar buff and would acquire very expensive and rare cigars from around the world. From time to time he would drop by my office and get me to try one out, explaining the origin and history of the company making it. These made a nice change from my usual fare of Tampa Nuggets and, when I had a good month, Dutch Masters.

One evening he called and invited Ree and me to a cheese party. I had never been to a cheese party, but I reasoned he would have some very fine hoop cheese with maybe a little Swiss to go along with it. Also, perhaps some really fine crackers to boot. For some reason, (kids?) Ree was delayed being able to go, so I went on over early. He had assembled ten or twelve friends. He explained the origin of the very rare cheese he was about to serve. He explained that this cheese was made up in small batches in Switzerland and placed deep inside some particular caves to cure or mature…I'm not sure which. It had to stay there exactly six months to the day and then be rushed to the customers by plane so that it could be eaten within six days. This meant that you had to place your order that far ahead. This whole procedure impressed this middle-class crowd very much. How nice it was to be rich!

Then he placed a plate of this cheese in front of each of us at the table. No crackers. In order not to upset you, I will not describe what this plate of cheese looked like, much less what it smelled like. I believed then and I believe now that this particular batch of cheese was either placed in the wrong cave too long or was a few days overdue at Bob's house, or some or all of the above. Anyway, he stood proudly by watching us consume our platter. I will say right now that I did our family proud in that I ate all of it and I gave every indication of enjoying it. (I was not about to cause my fine cigar source to dry up!) However, I will admit to being slightly ill by the time I finished.

About this time Ree arrived and sat down in her place beside me and received her plate full. One look and one whiff and her plate full of cheese magically changed from her plate to mine, right before my very eyes. This just wasn't an appropriate time for us to have an argument over this…she had me. So, despite being slightly ill, I had to consume Ree's plate too. I vowed right then and there that someday I would get her back for that little sleight of hand. She even sweetly told him just how much she enjoyed the meal as we left. It took me a long time, but one night in Beirut I got my chance and got her back in spades.

Along toward the end of our tour, we began to get visits from my old buddy from the 22nd Infantry Regiment days and our friend from the Heidelberg days, Autrey Maroun. He was now stationed in Washington and assigned to the Personnel (G-1) division of the Department of Army. Autrey had been a bachelor most of his life and was revered by all the wives for his chivalrous behavior toward women. He was, however, now married and his wife, named Amy, always came along on his visits. His purpose in visiting us on a regular basis, aside from our long friendship, was to get me to agree to being assigned as his Executive Officer in the manpower Division when I left the War College.

Much as I liked Autrey personally, I had never worked for him, and I was not sure I wanted to. Autrey, while a fine officer, had quite a reputation as a martinet and a stickler for all the rules. He was also somewhat volatile and could easily get upset when he was displeased. But, in time, after several visits, I decided I couldn't very well refuse. and it might well be better than being left to the tender mercies of some anonymous assignment office. So, sure enough, being in G-1 when assignments were made, Autrey was able to make the arrangement. So this is why my next assignment came about. I was due to report on the 16th of July, 1955.

So, our very pleasant and rather uneventful tour of one year at the Army War College came to an end, and we took a long leave at home. [54]

Edwards' Military Occupational Specialty was changed to 2025 (Major Departmental Unit Director or Chairman) on September 16, 1955.

He served in the Pentagon from 1955 to 1960.

In 1956 he attended and completed a three week Command Management Course at the Command Management School at Fort Belvoir, Virginia.

On September 1, 1956 Edwards became a member of the

General Staff Corps and thus became entitled to wear the General Staff Badge.

In 1960, he attended the United States Army Strategic Intelligence School at Fort Holabird, Baltimore, Maryland where he completed an eighteen week Attaché Course.

Edwards was the United States Military Attaché at the United States Embassy at Beirut, Lebanon from June 10, 1960 to June 18, 1962.

On July 1, 1960, Edwards was promoted to the permanent rank of Lieutenant Colonel in the Regular Army.

For his service as Attaché in Lebanon, Edwards received letters of commendation from the Adjutant General's Office, from the Army Chief of Staff for Intelligence, from the Assistant Secretary of State and from the Secretary of the Army.

Earl W. Edwards was retired from the Army with the rank of Colonel on July 31, 1962 at Fort Hamilton, New York.

He spent a total of 23 years, 2 months, and 9 days in Army service, which included his time as a Reserve officer. Of that total time 6 years, 8 months, and 14 days were served overseas.

POSTWAR POSTSCRIPT

After the 22nd Infantry Regiment was inactivated, Edwards still maintained close ties with fellow members of the Regiment.

In 1946, Edwards stopped in twice to see Colonel Charles T. Lanham, his former commanding officer in the Regiment, at Lanham's office in Washington. The first time, in March, Lanham was in Europe on special assignment. The second time on April 17, Edwards did meet with Lanham.

In either late 1949 or early 1950, Edwards and Lanham met

again, this time in Germany. Lanham was assigned to head the U.S. Military Assistance Advisory Group for Belgium and Luxembourg in 1949. Edwards was at the time stationed in Heidelberg, Germany and Lanham visited him there. In a letter Edwards wrote many years later, he recalled the meeting with Lanham in Germany:

When he was assigned to head MAAG Belgium (I think) he stopped in Heidelberg, Germany on the way and asked me to go with him. I was unable to agree to this assignment for a number of reasons but, due to the circumstances, I was unable to clearly explain it to him. I was always sorry about that because I think it hurt his feelings. [55]

Glenn D. Walker and his wife Margaret were close friends of Earl and Ree Edwards and they visited often. Edwards and Walker had become close friends beginning in 1942 when Walker was assigned to the 22nd Infantry at Camp Gordon, Georgia. They remained close friends for life.

Autrey Maroun who had served with Edwards in the 22nd Infantry in its early days before the war, twice worked on Edwards' behalf in getting him good assignments in the Army, once in postwar Germany and then later in getting Edwards an assignment in the Pentagon in the 1950's. Maroon retired as a Major General in 1972.

Bill Boice, who had served as the Regimental Chaplain during the war, together with Edwards and other officers formed the 22nd Infantry Officer's Association which evolved into the 22nd Infantry Association. Edwards attended many of the reunions of that organization and thus kept in touch with many of his fellow soldiers from the Regiment. Edwards was a member of that organization for fifty years.

In the early 1990's, under the direction of John F. Ruggles, who had commanded the Regiment in the final days of WWII, the Association opened its ranks to veterans of the 22nd Infantry Regiment from other periods of time, not just WWII. As the WWII veterans advanced in age, it became apparent the Association needed to turn over the running of its affairs to a younger generation.

John F. Ruggles asked Robert "Bob" Babcock, a veteran of 1st Battalion 22nd Infantry in Vietnam to take charge of the Association in order to move it into a new generation. Under Bob Babcock's direction, the Association then evolved into the 22nd Infantry Regiment Society. At the reunion of the Association in 1996 at Kissemee, Florida, for the first time veterans of the 22nd Infantry from other eras instead of just the World War II era were invited to attend the reunion. Earl Edwards, representing the WWII membership, for the first time passed the "Ruggles Torch" a trophy which symbolizes the passing of the leadership and the continuing of the traditions of the Regimental history to J. Bradley Hull, a veteran of the 2nd Battalion 22nd Infantry in Vietnam:

Finally, and definitely not least, **a new tradition was started with the presentation of the "Ruggles Torch." Tom Reid** had the idea to symbolize the passing of the leadership and continuation of the tradition of our Society by donating a trophy replica of a torch. **Earl "Lum" Edwards** represented the World War II membership and said a few words before he passed the torch to **Brad Hull** who represented the Vietnam and present day veterans. Brad accepted the torch on behalf of all the new and future members of our Society. The torch sat on the podium during the banquet. In the future, the torch will sit on the desk of the president during the year, to remind him of the responsibility he has to perpetuate this great Regimental Society. At all future reunions, the Ruggles Torch will sit in a position of prominence on the speaker's platform in all business meetings and the formal banquet. [56]

Another officer Edwards remained close friends with was Robert Brank Mclean from Shelbyville, Tennessee. Mclean had been a Liaison Officer between the 22nd Infantry and 4th Division Headquarters and ended the war as S-1 Officer or Adjutant for 2nd Battalion 22nd Infantry. When the remains of John Dowdy, the Commander of 1st Battalion 22nd Infantry who had been killed during the Regiment's first penetration into Germany in September 1944 were returned to the United States in 1949, Edwards wrote to Mclean, who wished to attend the funeral service. When Edwards died, Mclean's widow made a monetary donation to the 22nd Infantry Regiment Society, specifically to be used to honor the memory of Earl Edwards. That donation was used to create the "Lum" Edwards memorial gavel.

ERNEST HEMINGWAY

In late July 1944, the author and correspondent Ernest Hemingway attached himself to the 22nd Infantry Regiment. From July into December 1944, Hemingway stayed with the 22nd except for a few forays he made in search of other adventures. He always returned to the Regiment after the side trips he made elsewhere. Though Hemingway visited and mingled with several of the line Companies of the 22nd Infantry, he mostly associated with the Regimental leadership at the Regimental Command Post. Hemingway became personal friends with Colonel Buck Lanham the Regimental Commander, Lieutenant Colonel John Ruggles, the Regimental Executive Officer, and Lieutenant Colonel Earl Edwards, the Regimental Operations Officer. He also became a personal friend of Captain Don Warner, who took over command of Company A during the Battle of the Hurtgen Forest.

When the 22nd Infantry Officer's Association was formed, Hemingway was made a life member of the Association.

No confirmation could be found that Hemingway ever attended any 22nd Infantry Association reunion, but he always held the Regiment in high esteem:

He loved the 22nd Infantry and spent a great deal of time with various letter companies. He was always on line when we jumped off in the attack…He loved a fighting man and made no bones about this love he had for the 22nd, its men, and its officers. [57]

Each year, as long as he lived, he sent an ample supply of sumptuous baked turkeys to all 22 Infantry conventions, thanking the members in attendance for a job well done. Upon his passing, his son-in-law, a retired artillery colonel, continued this tradition—THE HEMINGWAY TURKEY!! [58]

The tradition of the Hemingway Turkey continues to the present day as a cooked turkey is consumed at every 22nd Infantry Regiment Society reunion to remember and honor the tradition started by Ernest Hemingway.

In 1948, Hemingway began work on his novel *Across the River and Into the Trees* in which the central character is an American officer during World War II. That officer is said to be modeled after Colonel Charles T. Lanham who commanded the 22nd Infantry Regiment. Without a doubt, Hemingway drew upon his wartime experiences with the 22nd Infantry Regiment as background material for the novel.

Hemingway and Edwards kept in touch after the war. Hemingway wrote to Edwards asking Edwards to compile a list of every place the 22nd Infantry had been during the war and the dates of such. As Operations Officer for the Regiment during the war, Edwards was eminently qualified to compile such a list. Hemingway also asked Edwards to write a summary of the actions of the 22nd Infantry during the Battle of the Hürtgen Forest, during which Hemingway had stayed with the Regiment in the front lines.

Don Warner had a memory of an incident that occurred between Edwards and Hemingway after the war and related it in a letter to Edwards' nephew John King:

"When your uncle, Colonel "Lum" Edwards was stationed at the Pentagon, another officer was aware of the close friendship existing between him and Hemingway, and imposed on this friendship in order to acquire Hemingway's autograph. This put Lum in a terrible position because Hemingway did not sign or give autographs. With pressure and great embarrassment, Lum, upon being pressured, sent another letter requesting same. Hemingway honored Lum's second request and on a single sheet of paper signed his name at the top and at the bottom included a very short note requesting never do this again. (This was the beginning of the ice age).

"On a visit to Lum & Ree in Tchula, many years after the fact of the autograph request, Lum related the above to me, confessing his deep embarrassment over this entire situation and deeply regretting pursuing same." [59]

At one time, Edwards wrote an outline for a speech about Ernest Hemingway. In that outline, Edwards described Hemingway as being physically strong, burley, and handsome. He said that Hemingway was shy and quiet with strangers, but talkative with friends, that he was actually a private person who was "informal with no put on, no stuffiness, and no sham." Edwards continued by saying that Hemingway liked all kinds of people and asked people to call him Ernie. Edwards said Hemingway was loyal and generous to his friends, that he had little money, but paid for everything at dinners in Paris and at conventions, helped friends in trouble, and had personally helped Edwards when Edwards had suffered an illness in combat. Edwards also noted that Hemingway had a wild side, drank to excess, had been married four times, and came to a bad end. [60]

In the early 1960's, when the author Carlos Baker began work on his biography of Hemingway, he reached out to Edwards for information on Hemingway during World War II. Edwards was at that time the military attaché to Lebanon. Edwards replied with much of the same descriptions of Hemingway as above and regretted that he could not be of much help. Baker replied that, on the contrary, Edwards had helped very much.

16. CIVILIAN EDUCATION AND MILITARY SCHOOLING				
SCHOOL	MAJOR OR COURSE	DURAT.	COMP.	YEAR
Mississippi St,Miss(Col)	AgriculturalEng	4yrs	BS	39
Infantry	Communications	12wks	Yes	41
Infantry	Rifle&HvyWpns	12wks	Yes	42
Infantry	Advanced	12wks	Yes	43
Engineer	ROTC Orient C.	1 wk	Yes	
Infantry	PMS&T	3 wks	Yes	46
TIS	Adv - Equiv	36wks	Yes	47
C&GSC	Basic - Equiv	16wks	Yes	47
C&GSC	15th Gen Staff	9 wks	Yes	43
C&GSC	Regular	10mos	Yes	54
ARWC	Regular	10mos	Yes	55
Comd Mgmt Sch	Comd Mgmt Crse	3wk	Yes	56
USA Strat Intel Sch	Attache Crse	18wks	Yes	60

That portion of the Army Service Record of Earl W. Edwardsshowing his education history. Note the record indicates he completed two courses pertaining to R.O.T.C. The first is on the line marked for the Engineer School and is an R.O.T.C. Orientation Course (ROTC Orient C.) which lasted for one week. The date is marked out but is most likely 1946. The second is on the line immediately below that and is a Professor of Military Science and Tactics Course (PMS&T) at the Infantry School which lasted for three weeks and has a date of completion of 1946. These courses would have been necessary for Edwards to be accredited as an R.O.T.C. instructor which was his first assignment after the deactivation of the 22nd Infantry. On March 31, 1946 Earl W. Edwards' Military Occupational Specialty was changed to 2517 Professor of Military Science and Tactics.

Michael D. Belis

The R.O.T.C. staff at the University of Florida 1946-1947 Lieutenant Colonel Earl Edwards is sitting in the front row on the far left.

Lieutenant Colonel Earl W. Edwards as Head of the
Infantry Section and Executive Officer of the R.O.T.C.
program at the University of Florida 1946-1947.

Lieutenant Colonel Earl W. Edwards in Germany circa 1950-1952

Lieutenant Colonel Earl W. Edwards touring Heidelberg, Germany

Earl "Lum" Edwards on the left, Glenn D. Walker on the right. Photo taken at Fort Leavenworth, Kansas 1952-1953. They remained friends for life.

WW II

~~AIRP~~ ...

SEPT 1939 TO AUG 1945
2174 DAYS - ALMOST 6 YRS

46 MILLION ...-CIVILIANS KILLED
... WOMEN + CHILDREN

USE OF AIRPLANE IN WAR - JETS
  STRATEGIC BOMBER
USE OF ATOMIC WEAPONS

HITLER-GERMANY STARTED WAR

JAPAN BEGAN EXPANSION

WHAT CAUSED WW II

GERMANY ANGRY OVER TERMS
OF WWI - ANXIOUS TO EXPAND

JAPAN, ANXIOUS TO EXPAND
  NO OIL - NO COAL - SHORT FOOD
  ANGRY OVER RESTRICTIONS

US NOT INITIALLY INVOLVED

HOW DID WE GET INVOLVED

JAPAN FIRST - GERMAN SECOND

WHY FIGHT JAPAN
  THEIR AMBITIONS INCLUDED
  US TERRITORY

WHY GERMANY
  FEAR US WOULD BE LEFT
ISOLATED IN HOSTILE WORLD

US STRATEGY
  BEAT GERMANY FIRST
AS MOST DANGEROUS TO US
  HOLD OR SLOW UP JAPAN
UNTIL WE BEAT GERMANY
  THEN FINISH OFF JAPAN

UNCONDITIONAL SURRENDER

POLICY AFTER THE WAR

NOT TO MAKE MISTAKES
MADE AFTER WWI -
IMPOSE REASONABLE TERMS
AND TRY TO BRING BOTH
GERMANY + JAPAN BACK
INTO PEACEFUL RELATIONS
WITH REST OF WORLD

Outline for a speech written in Edwards' handwriting. Date is
unrecorded but is most likely for a speech he gave at the Command
& General Staff College or at the Army War College.

```
                                    DEPARTMENT OF THE ARMY,
SPECIAL ORDERS)                     WASHINGTON 25, D.C., 31 January 1955
NUMBER     21)

                        E-X-T-R-A-C-T

  *            *            *            *            *

    11.  DP announcement is made of the temp promotion and commissioning
of the fol-named officers in the Army of the United States under the
provisions of subsec 515 (c) of the Officer Personnel Act of 1947 in gr
and with date of rank as indicated:

            LT COL TO COL WITH RANK FR 31 JAN 1955

              *            *            *

            PAUL T. CLIFFORD 022135 Inf

              *            *            *

            RICHARD J. DARNELL 024159 Inf

              *            *            *

            EARL W. EDWARDS 023502 Inf

              *            *            *
```

The orders promoting Earl W. Edwards to Colonel in the
temporary organization of the Army of the United States
(AUS). Edwards' name is underlined at the bottom.

Photo of Army officers with Earl W. Edwards standing in the second row second from the right. Date and location is unrecorded, but photo can be dated by the olive drab wool serge uniforms and the wearing of the overseas bars on the lower right sleeve. Those attributes date the photo as 1953-1959.

Left to right: Colonel Earl W. Edwards, Ree Edwards, Greek
Ambassador. Photo most likely taken at the U.S. Embassy in
Beirut circa 1960-1962 since Edwards is wearing the gold
aiguillette of a military attaché over his left shoulder.

The decorations of Earl W. Edwards. Top: Combat Infantryman Badge. Center left to right: Silver Star Medal, Bronze Star Medal, American Defense Service Medal, American Campaign Medal, European-African-Middle Eastern Campaign Medal with arrowhead and silver service star, World War II Victory Medal, Army of Occupation Medal with Germany clasp, National Defense Service Medal, French Croix de Guerre with palm. Bottom left to right: Three overseas service bars, General Staff Badge, (above) Presidential Unit Citation with oak leaf cluster, (below) Scabbard and Blade ribbon. He was also awarded the Belgian Fourragere.

Earl Edwards second from the right visiting with old comrades of the 22nd Infantry at a 22nd Infantry Association reunion. The photo was marked as "Story Telling."

Captain Robert B. Mclean Awarded the Silver Star Medal. Like many
of the officers who served in the 22nd Infantry during the war, Mclean
continued service afterward in the National Guard. Known by his
middle name of Brank, he and Edwards remained friends for life.

The cover of *Journal of the Hurtgen Forest 22d Inf 16 Nov 44—3 Dec 44*, the 93 page report on the activities of the 22nd Infantry during the Battle of the Hürtgen Forest as compiled and written by Earl Edwards to fulfill a request made by Ernest Hemingway.

his journal made up of action after action reports,coming

hrough the S-3 section commanded by Lt. Col. Earl W. Edwards,

egimental operations officer, is a unique accounting of events

ut together in book form, at the request of Ernest Hemingway,

ar correspondent.

his undertaking ranks high as a rare masterpiece, a brilliant

em and a work of art amoung literary and military historians.

ol. Earl W. Edwards and staff are to be commended for preserv-

ng this segment of regimental history for future generations.

Deeds Not Words   Able Red Six

The introduction to *Journal of the Hurtgen Forest 22d Inf 16
Nov 44 — 3 Dec 44* was written by Captain Don Warner as
Commanding Officer of Company A (Able Red Six.)

Michael D. Belis

Colonel Earl W. Edwards
U.S.Army Attaché
American Embassy
Beirut, Lebanan

Dear Colonel Edwards,

    I cannot thank you adequately for your prompt,
full, and generous response to my request for information
on Ernest Hemingway in the European War. In spite of
your protestation to the contrary, I found your letter
both eloquent, filled with readings of his character and
habits which are exactly the sort of information I need.

    Thank you very much indeed. If you ever see my old friend,
Colonel Bill Eddy, who lives in Beirut, please remember me to him.

              Yours very sincerely,

              Carlos Baker

              Carlos Baker

Letter written to Earl Edwards by Carlos Baker, Hemingway's biographer

4. The Later Years

After leaving military service, Edwards spent thirteen years in a career in education.

In 1965, Edwards was one of the founding members of the Cruger-Tchula Academy, a private High School in Holmes County, Mississippi. From 1965 to 1973, Edwards was the Headmaster and Principal at the Academy. Edwards was affectionately known at the Academy as "The Colonel". The yearbook for the Cruger-Tchula Academy was called "The Colonel" and the school team name for sports was "The Colonels."

In 1973, Edwards left the Academy and took a position with the Mississippi Private School Association as Executive Secretary. He remained in this capacity for five years.

In 1978, Edwards retired to spend time with his family. Earl and Ree had three children. Patricia Lyn was born on July 29, 1950, Bruce Alan was born on July 20, 1952 and Mark Kent was born on September 15, 1957.

Though Earl Edwards was called "colonel" ever since he was promoted to Lieutenant Colonel in 1944 and was known in his post-Army service as "Colonel" because he retired as a Colonel, his wife Ree always referred to him as "the Private."

From the earliest days of his childhood, Edwards liked to read. He was a student of history, especially the American Civil War, and his specialty in that area was the study of Ulysses S. Grant. He continued to be an avid reader for all his life.

In his later years, Edwards continued to take an active part in the 22nd Infantry Regiment Society. He was also contacted by authors in search of information and details for their works on aspects of the history of the 22nd Infantry Regiment. In the early 1990's, when Edward G. Miller was doing research for his book on the Battles in the Hürtgen Forest, he contacted Edwards. Edwards wrote several letters to Miller in which he described many aspects of the 22nd Infantry Regiment during World War II. In those letters, Edwards recorded important history of the Regiment with the charisma of his personal insight.

One of those letters to Miller was a four page profile of Colonel Charles T. Lanham who commanded the 22nd Infantry Regiment during the war. A second letter gave Edwards' thoughtful analysis of the 22nd Infantry Regiment during the war:

Dear Major Miller,

You made reference to the "close-knit" character of the 22nd Infantry Regiment and asked for possible reasons.

The 22nd Infantry (or Double Deucers) as we liked to call ourselves, was a very close-knit unit—very much like a family. This feeling of closeness was evident between the officers as well as between officers and enlisted men. I also do not believe there was ever any gulf between command and staff on the one hand and line officers on the other. We could, like siblings in a family, quarrel and argue, but our respect and affection for each other always held us together. I joined the regiment in

July, 1940 and left it when it was disbanded at Camp Butner in 1946. It is my belief that this feeling began back in the early years and gradually grew as the years went by. The very strong interest shown by both officers and enlisted men for the 22nd Infantry Association (now Society) is evidence to that. The attendance and interest in our conventions over the years has been amazing and seems to get stronger every year. This year's convention may see the largest attendance ever despite the gradual decline in the health of so many of us. Officers and enlisted men attend from all of those years. It is my understanding that we are the only unit with a regimental association.

I have often heard Lt. Gen. Glenn Walker comment on this aspect. Glenn joined the regiment as a Second Lieutenant and left it when he was wounded as a battalion commander. Glenn served on to become a Lieutenant General, so he has a wide range of experience and assignments to speak from.

You ask the question, "Why?" This is hard to say. Why any particular team, be it a company, athletic team, military unit, etc. welds together as a cohesive, close-knit unit is open to argument. If any coach or company president could find that secret he could make a fortune. Mostly, I think it is a lucky combination of individuals who happen to be blessed with the right kind of leadership.

Looking at the leadership, I would say we had a series of fine, well-respected commanding officers, starting with Colonel Albert S. Peake in the 40's followed by Col. "Daddy" Weems and on into early combat with Col. Tribolet.

Of these I think most of us would give great credit to Col. Tribolet. Most of us would agree that he built the regiment. He was a quiet, gentle father figure, much beloved and respected by all of us. None of us wanted to be seen by him as unmannerly or disrespectful nor did we want to hurt his feelings in any way—so we learned to get along.

I remember one time when Guy DeYoung, Regimental S-4, and I clashed over a supply problem. Since we were unable to agree, I suggested we take it to Col. Tribolet for resolution. Guy thought a minute and said, "Now how would we look to Col. Trib being unable to solve a little problem like this?" I agreed and we quickly found a solution. What he thought about us was important.

Colonel Tribolet couldn't handle the regiment in combat, and we all understood that. Colonel Lanham, and later Colonel Ruggles, fully understood the "family" nature of the regiment and wisely played to that strength in their command style.

Then it did seem to me that we were more than fortunate in the number of fine enlisted men and officers we wound up with. Lucky, too, in that we had so few of the really disruptive types.

In very large part the officers came from the South. The Citadel, Mississippi State, LSU, University of Alabama, etc. all had strong contingents. The University of Maryland was about as far north as we got.

The enlisted men were, as I remember it, mostly from the North. Exactly why we developed such a fine relationship is beyond me. There certainly was a strong bond between them that holds to this day. I can never remember that we had a discipline problem of any kind. When we started the association in 1946, we assumed that only officers would be interested so we started out in that way. We soon learned better.

I wish I could give you a better explanation, but I can't.

I still feel chance played a big part as it does in so many things.

Sincerely,
Lum Edwards [61]

Edwards wrote another letter to Miller the day after he wrote the letter above and in this letter Edwards expanded upon the theme in his previous letter:

Tchula, Mississippi May 11, 1994

Dear Major Miller,

I do not have much of a recollection of an attack on the regimental CP when Ernest Hemingway entered the battle. I do remember one time when we by-passed a German unit and, unfortunately, placed our CP close to it. One of our units had to be brought back to clear them out. I remember Carl Warren, CO of the 44th FA, calling me up to explain that they would be firing very close to the CP in support of this action. He said for us not to worry—that the range had been carefully calibrated. He also gave the approximate time the firing would take place. I passed this word around the CP and told everyone not to worry. Unfortunately, I chose this time to attend an open latrine just beside the CP. As the artillery started falling, some of it landed in the CP and the rest of it too close for comfort. There was a great deal of raucous laughter as I ran for cover across the CP unable to get my bunglesome clothes up.

I also remember Brank McLean, at that time acting as Division Officer, trying to calm the fears of another officer who

189

was near a mental breakdown by explaining that he shouldn't worry since this was friendly fire. The officer, in great distress and with much emotion and a certain logic, replied, "Friendly fire will kill you just like any other kind."

I do not know if this was the action you have reference to or if Hemingway was involved. I am fairly sure, based on some other references that this occurred in the Hurtgen.

To the best of my knowledge, we did not receive any briefings or special training prior to entering the Hurtgen. We rarely had time to do any kind of training since we were almost always in combat or moving to be in contact.

I do not have any information about Capt. Faulkner — I'm sure someone at the 22nd Infantry Convention in May could help you on that.

I would like to add a little more reference the close-knit character of the 22nd Infantry. Another possible basis occurred to me after I wrote my last letter and that is that almost all of the battalion commanders, battalion executives, regimental staff officers, and, in many cases, company commanders joined the 22nd in the early 40's and grew up together. I could name a long list but hesitate to do so that I might leave someone out. It just seems natural to me that this would add to the cohesiveness of the unit and explain their great willingness to give support and cooperation to each other in combat.

Another aspect that I would like to emphasize more is the wonderful relationship between the officers and enlisted men. When I read stories about Vietnam, about some of the discipline problems, perhaps exaggerated, I get a warm glow when I think back on my service in the 22nd Infantry.

I cannot remember any kind of disciplinary problem occurring before or during combat. We had a closeness and camara-

derie that is a little hard to explain. We all just seemed to like and respect each other.

I remember being at one 22nd Infantry Convention and overhearing a group of enlisted men discussing their officers. As each name came up it received full approval and at the end of the discussion one man said, "We were just lucky to have such a great set of officers."

It is also something of a surprise to me that so many wives continue to attend the convention even after their husbands have passed away.

I will mail you a summary of my service career along with a photo of the time soon.

Sincerely,
Earl W. Edwards [62]

When Command Sergeant Major Robert S. Rush published his work on the 22nd Infantry Regiment in the Battle of the Hürtgen Forest *Paschendale with Treebursts* in September 1996, one of his research sources was Earl W. Edwards. Rush included an Afterword to the work written by Earl Edwards. When the work was published in 2001 by the University Press of Kansas as *Hell in Hürtgen Forest* the same Afterword was included. In it Edwards related the sadness he felt after the battle:

The author properly points out that while a steady stream of replacements kept the numbers up, the erosion in the ranks of experienced leaders, commissioned as well as non-commissioned, led gradually to the virtual destruction of a great regiment. I had been with the regiment since 1940 and knew just about everyone. I remember standing on the side of the road going into the Hürtgen and being waved at by soldiers from every truck. When we left the

Hürtgen, I again stood on the side of the road; but this time there were not as many trucks and very few soldiers waving. Everyone else was a casualty. [63]

Edwards continued to enjoy being active with his family and working with the 22nd Infantry Regiment Association (Society) during his retirement.

In the spring of 1996, he attended his last 22nd Infantry Regiment Society reunion at Kissimmee, Florida.

On August 9, 1997 Earl William "Lum" Edwards died of leukemia at the University of Mississippi Medical Center in Jackson, Mississippi.

Services were held on August 12, 1997 in Greenwood, Mississippi and he was buried in Pinecrest Cemetery in Tchula, Holmes County, Mississippi.

The obituary for Earl Edwards stated that he was 79, and a retired U.S. Army Colonel from Tchula, Mississippi. Services for him were held at 10:00 a.m. on Tuesday, August 12, 1997 at Wilson and Knight Funeral Home in Greenwood, Mississippi, with Reverend Weldon Greer officiating. The obituary stated that he was a Cruger native, attended Cruger High School, and graduated from Mississippi State University. It mentioned he served in the Army from 1939-1962, was a World War II veteran having served with the 22nd Infantry Regiment. He was a former headmaster of Cruger-Tchula Academy and former Executive Secretary of the Mississippi Private School Association. He was a member of the Cruger Independent Methodist Church. Survivors included his wife, his daughter, two sons, two brothers, two sisters, and six grandchildren.

Active pallbearers were Vemon Lehman, Hoss Simon, David Flemming, Steve Flemming, Bobby Hugh Spivey, Bill Muriagh. Danny Edwards, and Al Barfield.

Honorary pallbearers were Don Warner, Lewis Massey, the stewards of Cruger Independent Methodist Church, and all 22nd infantry veterans.

The grave marker for Earl W. Edwards is inscribed with

DEEDS NOT WORDS

the official motto of the 22nd Infantry Regiment.

Michael D. Belis

The Cruger-Tchula Academy, a private High School
which existed from 1965 to 2001.

194

We, the students of Cruger-Tchula Academy, dedicate our first edition of the Colonel to one whose willingness to listen and offer his help has played an important part in this school year. His enthusiasm and efficiency in his work has contributed a great deal toward the successful operation of our school. His devotion to duty during these critical and pressing times has remained strong and steadfast. His loud, booming laugh has added a touch of cheerfulness to all his undertakings. Because of these reasons, we the student body dedicate this edition to our principal.

Mr. Earl W. Edwards

Dedication

The dedication page for the first edition of the Cruger-Tchula Academy yearbook the "Colonel" indicating that the yearbook was dedicated to Earl W. Edwards.

Earl W. Edwards speaking before a meeting of the Mississippi
Private School Association. Date and location unknown.

Earl William Edwards

Endnotes

1. Letter from and conversation with Gladys Edwards King, sister of Earl Edwards, May 24, 2013.

2. Earl W. Edwards memoirs, unpublished manuscript. Courtesy of John and Gladys King.

3. Ibid.

4. Ibid.

5. Ibid. Morgan G. Roseborough would retire a Major General in 1975.

6. Ibid.

7. Ibid.

8. Ibid.

8a. Ibid.

9. *THIS IS MY STORY* unpublished manuscript by Charles A. Mastro. (Mastro was a member of Company E 2nd Battalion 22nd Infantry 1940-1945. Courtesy of John and Gladys King.

10. *WAR STORIES Utah Beach to Pleiku* by Robert O. Babcock, Saint John's Press, Baton Rouge, Louisiana, 2001 pp40.

11. Earl Edwards estate, courtesy of John and Gladys King.

12. Earl W. Edwards memoirs, unpublished manuscript. Courtesy of John and Gladys King.

13. Ibid.

14. *THIS IS MY STORY* unpublished manuscript by Charles A. Mastro. Courtesy of John and Gladys King.

15. *The Road from Caledonia to Canisy, One man's Journey from Home Through World War II And Back* by Bill Andrews 2003 pp55-56.

16. Ibid pp66.

17. Ibid pp68.

18. Ibid pp120.

19. Ibid pp120-121.

20. *WAR STORIES Utah Beach to Pleiku* by Robert O. Babcock, Saint John's Press, Baton Rouge, Louisiana, 2001 pp163.

21. *History of the Twenty-Second United States Infantry in World War II* by William S. Boice 1959 pp25.

22. Ibid pp26.

23. *WAR STORIES Utah Beach to Pleiku* by Robert O. Babcock, Saint John's Press, Baton Rouge, Louisiana, 2001 pp159-160.

24. Swede Henley Diary 22nd Infantry Regiment 4th Infantry Division WWII compiled and printed by Tommy Harrison 1973 pp7.

25. *History of the Twenty-Second United States Infantry in World War II* by William S. Boice 1959 pp26.

26. *The Road from Caledonia to Canisy, One man's Journey from Home Through World War II And Back* by Bill Andrews 2003 pp150.

27. Ibid pp150-151.

28. Ibid pp151.

29. *History of the Twenty-Second United States Infantry in World War II* by William S. Boice 1959 pp27.

30. Ibid pp28.

31. Swede Henley Diary 22nd Infantry Regiment 4th Infantry Division WWII compiled and printed by Tommy Harrison 1973 pp7.

32. *History of the Twenty-Second United States Infantry in World War II* by William S. Boice 1959 pp28.

33. *The Road from Caledonia to Canisy, One man's Journey from Home Through World War II And Back* by Bill Andrews 2003 pp152-153.

34. Ibid pp153.

35. Ibid pp153-154.

36. 22nd Infantry Regiment Society Newsletter.

36ª. *Ernest Hemingway: A Life Story* by Carlos Baker Charles Scribner's Sons: NY 1969 pp. 402-403.

37. Earl W. Edwards memoirs, unpublished manuscript. Courtesy of John and Gladys King.

37ª. *Ernest Hemingway: A Life Story* by Carlos Baker Charles Scribner's Sons: NY 1969 pp. 433.

38. Letter from Don Warner to John King undated. Courtesy of John and Gladys King.

39. Ibid.

40. 22nd Infantry Regiment yearbook published by the 22nd Infantry Regiment, 4th Division Printed by *Army & Navy Publishing Company*, Baton Rouge, Louisiana 1946

41. Earl W. Edwards memoirs, unpublished manuscript. Courtesy of John and Gladys King.

42. Letter from Don Warner to John King July 19, 2006. Courtesy of John and Gladys King.

43. Movement Orders CKCB 370.5 (# 1269) ARMY SERVICE FORCES New York Port of Embarkation Camp Kilmer 11 July 1945.

44. Letter from Don Warner to John King July 19, 2006. Courtesy of John and Gladys King.

45. Earl W. Edwards memoirs, unpublished manuscript. Courtesy of John and Gladys King.

46. Telephone conversation with John King, the nephew of Earl Edwards 2017.

47. Earl W. Edwards memoirs, unpublished manuscript. Courtesy of John and Gladys King.

48. Letter from Don Warner to John King July 1, 2006. Courtesy of John and Gladys King.

49. Letter from Charles T. Lanham to Ernest Hemingway April 19, 1946. From the Ernest Hemingway Collection in the John F. Kennedy Library. Courtesy of John and Gladys King.

50. Earl W. Edwards memoirs, unpublished manuscript. Courtesy of John and Gladys King.

51. Ibid.

52. Ibid.

53. Ibid.

54. Ibid.

55. Letter from Earl Edwards to author Edward G. Miller October 1, 1993. Miller was conducting research for his book *A Dark and Bloody Ground* about the Battles of the Hürtgen Forest and had asked Edwards for descriptive information on Colonel Charles T. Lanham.

56. 22nd Infantry Regiment Society Newsletter Spring 1996. Courtesy of John and Gladys King.

57. Letter from Don Warner to John King July 25, 2006. Courtesy of John and Gladys King.

58. Letter from Don Warner to John King June 3, 2009. Courtesy of John and Gladys King.

59. Letter from Don Warner to John King June 3, 2009. Courtesy of John and Gladys King.

60. Handwritten outline of a speech about Ernest Hemingway written by Earl Edwards. Undated. Courtesy of John and Gladys King.

61. Letter from Earl W. Edwards to author Edward G. Miller May 10, 1994. Courtesy of John and Gladys King.

62. Letter from Earl W. Edwards to author Edward G. Miller May 11, 1994. Courtesy of John and Gladys King.

63. *Paschendale with Treebursts A History and Analysis of the 22nd Infantry Regiment During the Battle of the Hürtgen Forest 16 November through 3 December*

1944 by Robert S. Rush 1996 pp. 168. This same Afterword was included when Robert S. Rush expanded his work and it was published as *HELL in HüRTGEN FOREST The Ordeal & Triumph of an American Infantry Regiment* by Robert Sterling Rush University of Kansas Press 2001 pp. 348-349.

Illustrations — Credits

The "Lum" Edwards memorial gavel. Upon the death of Earl "Lum" Edwards a donation was sent to the 22nd Infantry Regiment Society for the express purpose of creating a permanent memorial to Edwards. The donation was sent in by Elizabeth Mclean, widow of Robert B. Mclean who had served as a Captain in the 22nd Infantry along with Edwards. Bob Babcock devised the idea of using the donation to create the "Lum" Edwards memorial gavel. The gavel is passed to each new President of the 22nd Infantry Regiment Society. Photo by Bob Babcock.

Photo of Lula and Homer Dee Edwards courtesy of John and Gladys King.

Photo of the Edwards children courtesy of John and Gladys King.

Photo of Earl Edwards in his father's store in Cruger, Mississippi courtesy of John and Gladys King.

Photo of Earl Edwards and ROTC students at Mississippi State

University courtesy of John and Gladys King.

Photo of Earl Edwards with motorcycle courtesy of John and Gladys King.

Formal portrait of Earl Edwards as a Lieutenant with the 22nd Infantry courtesy of John and Gladys King.

Portrait of John Dowdy as a Captain in the 22nd Infantry courtesy of Karen Scott the cousin of John Dowdy.

Company photo of Company B 22nd Infantry taken at Fort Benning circa June-October 1941 courtesy of John and Gladys King.

Photo of Earl Edwards from a Company photo of Company B 22nd Infantry taken at Fort Benning circa June-October 1941 courtesy of John and Gladys King.

Photo of Glenn D. Walker as a Major General in command of the 4th Infantry Division 1 August 1970 from U.S. Army Office of Public Affairs.

Photo of Earl Edwards in the 15th Command and General Staff Class 1943 courtesy of John and Gladys King.

Photo of Earl Edwards as a Major prior to D-Day. Most likely taken in England. Courtesy of John and Gladys King.

Photo of Field Marshall Bernard Montgomery arriving at the 22nd Infantry area. Exiting the car immediately behind Montgomery is Major General Raymond O. Barton Commander of the

4th Infantry Division. Courtesy of John and Gladys King.

Photo of Montgomery talking to Earl Edwards at 22nd Infantry area in England prior to D-Day. Also in the photo in the background to the immediate right of Montgomery is Brigadier General Henry A. Barber Jr. Assistant Commander of the 4th Infantry Division. Courtesy of John and Gladys King.

Photo of Combat Team 22 leaders in marshalling area in England just prior to loading on ships. Front row left to right: Colonel Hervey A. Tribolet Commanding Officer 22nd Infantry, Lieutenant Colonel John F. Ruggles Regimental Executive Officer 22nd Infantry, Lieutenant Colonel Arthur S. Teague Commanding Officer 3rd Battalion 22nd Infantry, Lieutenant Colonel Sewell M. Brumby Commanding Officer 1st Battalion 22nd Infantry. Rear row left to right: Lieutenant Colonel William A. Atson Commanding Officer 44th Field Artillery, Lieutenant Colonel Thomas A. Kenan S-3 Officer 22nd Infantry, Major Earl W. Edwards Commanding Officer 2nd Battalion 22nd Infantry. From the 22nd Infantry yearbook done in 1946. Published by the 22nd Infantry Regiment and printed by Army and Navy Publishing Company, Baton Rouge, Louisiana.

Photo of Earl Edwards looking out over battlefield from a captured German position courtesy of John and Gladys King.

Photo of James B. Burnside from the 22nd Infantry yearbook done in 1946. Published by the 22nd Infantry Regiment and printed by Army and Navy Publishing Company, Baton Rouge, Louisiana.

Photo of Joseph C. Rickerhauser from the 22nd Infantry year-book done in 1946. Published by the 22nd Infantry Regiment and printed by Army and Navy Publishing Company, Baton Rouge, Louisiana.

Photo of Lanham and Hemingway from 8mm film movies taken during the war by Clifford "Swede" Henley of the 22nd Infantry.

Photo of command post trailer used by Colonel Charles T. Lanham. Soldier sitting on steps of trailer is unknown and not believed to be Earl Edwards. Courtesy of John and Gladys King.

Photo of building in Zweifall used as 22nd Infantry command post from a video filmed by Arne Esser. Courtesy of John and Gladys King.

Photo of the area of Edwards' command post in the Hurtgen from a video filmed by Arne Esser. Courtesy of John and Gladys King.

Photo looking out from the area of Edwards' command post in the Hurtgen from a video filmed by Arne Esser. Courtesy of John and Gladys King.

Photo of Goforth, Edwards and Ruggles courtesy of John and Gladys King.

Photo of Edwards and Harrison with German car from 8mm film movies taken during the war by Clifford "Swede" Henley of the 22nd Infantry.

Photos of Soldbuch of Gefreiter Rudolf Sedlaczek. His rank is usually translated as Private First Class but sometimes as Lance Corporal. Sedlaczek appears to be in a German Air Force uniform and his issuing unit is difficult to determine from the scans. His Soldbuch however is for the German Army indicating that he most likely was impressed into an Army unit which was done often in the final stages of the war. Courtesy of John and Gladys King.

S-3 Periodic Report No. 202 of Combat Team 22 dated February 5, 1945 from the Journal Files of the 22nd Infantry for February 1945 at the National Archives scanned by John Tomawski.

Photo of Earl Edwards at Bad Mergentheim, Germany from 8mm film movies taken during the war by Clifford "Swede" Henley of the 22nd Infantry.

Photo of Thomas A. Kenan from the 22nd Infantry yearbook done in 1946. Published by the 22nd Infantry Regiment and printed by Army and Navy Publishing Company, Baton Rouge, Louisiana.

Photo of Earl Edwards as S-3 Operations Officer of the 22nd Infantry in 1946 from the 22nd Infantry yearbook done in 1946. Published by the 22nd Infantry Regiment and printed by Army and Navy Publishing Company, Baton Rouge, Louisiana.

Army Service Record of Earl W. Edwards courtesy of John and Gladys King.

Photo of Earl W. Edwards with R.O.T.C. Staff at University of Florida courtesy of John and Gladys King.

Photo of Earl W. Edwards as R.O.T.C. Instructor at University of Florida courtesy of John and Gladys King.

Photo of Earl W. Edwards in Germany circa 1950-1952 courtesy of John and Gladys King.

Photo of Earl W. Edwards at Heidelberg, Germany circa 1949-1952 courtesy of John and Gladys King.

Photo of Glenn D. Walker and Earl W. Edwards from post WWII courtesy of John and Gladys King. Identification of location and date by Lieutenant General Keith Walker, the son of Glenn D. Walker, as related in e-mail conversation with Michael Belis.

Handwritten notes by Earl Edwards for a speech about World War II courtesy of John and Gladys King.

Orders promoting Earl W. Edwards to Colonel courtesy of John and Gladys King.

Photo of Earl W. Edwards in group of officers. Date and location of photo is unrecorded. The uniform details date the photo as having been taken 1953-1959. Courtesy of John and Gladys King.

Photo of Colonel Earl W. Edwards, his wife Ree and the Greek Ambassador. Date and location of photo is unrecorded. Courtesy of John and Gladys King.

Graphic of Earl W. Edwards' decorations created by Michael Belis.

Photo of Earl Edwards and fellow veterans of the 22nd Infantry.

Date and location unknown. Courtesy of John and Gladys King.

Photo of Robert B. Mclean from the 22nd Infantry yearbook done in 1946. Published by the 22nd Infantry Regiment and printed by Army and Navy Publishing Company, Baton Rouge, Louisiana.

Scan of the front cover of the report on the activities of the 22nd Infantry in the Hürtgen Forest courtesy of John and Gladys King.

Letter from Carlos Baker to Earl Edwards February 12, 1962. Courtesy of John and Gladys King.

Photo of Cruger-Tchula Academy courtesy of John and Gladys King.

Scan of Dedication page of the *Colonel*, the yearbook of the Cruger-Tchula Academy courtesy of John and Gladys King.

Photo of Earl Edwards speaking at a Mississippi Private School Association meeting courtesy of John and Gladys King.

Photo of Earl W. Edwards courtesy of John and Gladys King.

Addendum

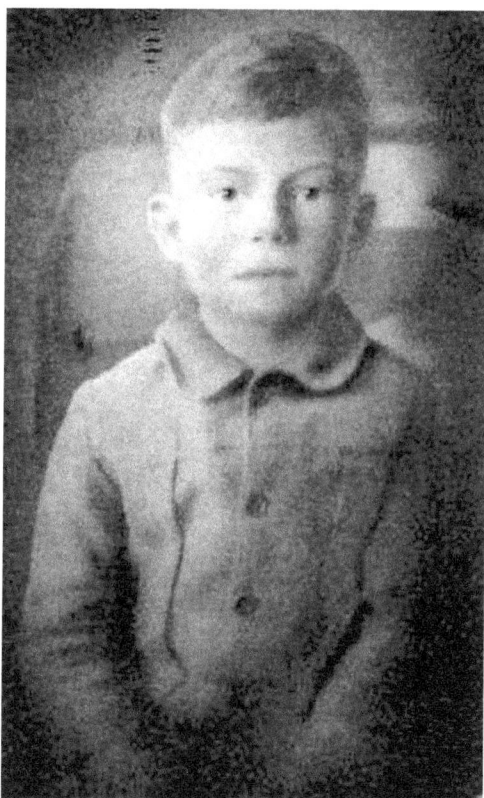

Earl W. "Lum" Edwards as a child. Age not recorded.
Courtesy of John and Gladys King.

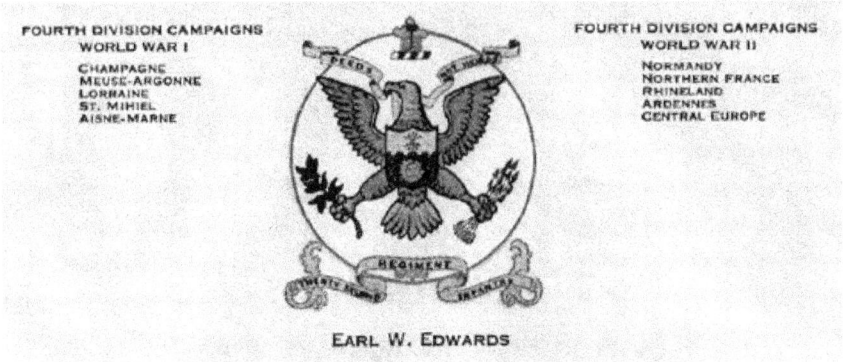

Letterhead of the personal stationery of Earl W. Edwards.
Courtesy of Karen Scott, cousin of John Dowdy.

The letterhead is constructed using the design from the 22nd Infantry Regimental Colors (flag) as the centerpiece. The official motto of the Regiment "Deeds Not Words" is in the scroll at the top.

The campaigns in which the 4th Infantry Division participated in both World Wars are given on either side.

Though the 22nd Infantry served in all campaigns of the 4th Infantry Division in World War II the Regiment was not part of the 4th Division in World War I.

Edwards used this stationery to write a letter to the mother of John Dowdy who was killed in action commanding 1st Battalion 22nd Infantry on September 16, 1944. In the letter dated April 2, 1949 Edwards informed Dowdy's mother that Edwards and his wife Ree would attend the re-burial services for John Dowdy whose remains were returned from Belgium and were about to be reinterred in Tifton, Georgia.

That letter may be viewed on the 1st Battalion 22nd Infantry website at: http://1-22infantry.org/commanders/dowdypers.htm

Photo of officers post World War II. Earl Edwards is on the far left. Date and location of photo is not recorded. Courtesy of John and Gladys King.

Edwards is wearing his Combat Infantryman Badge on his left breast and over his right shoulder he is wearing the Belgian Four-ragere. (The correct manner of wear of the Belgian Fourragere is over the left shoulder.)

Edwards and the officer second from the right are wearing their overseas service bars on their lower left sleeves.

The location of the overseas service bars on the left sleeve and Edwards wearing the Belgian Fourragere date this photo as having been taken 1946 — 1953.

Lieutenant General Walter Weible Deputy Chief of Staff for Operations & Administration. He was Earl Edwards' boss during Edwards' assignment to the Pentagon. He personally signed his best wishes to Earl Edwards on the above photograph.Courtesy of John and Gladys King.

Colonel Earl W. Edwards on the right with a foreign military
officer. Date and location are not recorded. Edwards is wearing the
General Staff Badge on his right breast which indicates the photo
was taken 1956 — 1962. Courtesy of John and Gladys King.

The personal medal display of Earl W. Edwards.
Courtesy of John and Gladys King.

Brothers and Sisters—The Edwards. Left to right: H.D. (Homer Dee), Fred, Gladys, Charline, Lum (Earl). Courtesy of John and Gladys King.

About the Author

Michael D. Belis served as a rifleman in the United States Army's 1st Battalion 22nd Infantry Regiment of the 4th Infantry Division in the Republic of Vietnam in 1970 and is a proud holder of the Combat Infantryman Badge. He is a member of the 4th Infantry Division Association and the 22nd Infantry Regiment Society.

From 1999-2014, he was a historical advisor to 1st Battalion 22nd Infantry while the Battalion was part of the 4th Infantry Division (Mechanized) based out of Fort Hood, Texas and then Fort Carson, Colorado, and contributed an extensive number of historical artifacts of the Regiment to the Battalion's Regimental Room during those years. Since 2014, he has been working in a historical advisory capacity to 2nd Battalion 22nd Infantry Regiment, 10th Mountain Division stationed at Fort Drum, New York, and has also contributed numerous artifacts of the Regiment to that Battalion for their displays at their home base.

He is the Historian for the 22nd Infantry Regiment Society and for over twenty years has written articles for the 22nd Infantry Regiment Society newsletter and for other projects by the Society. He has written articles for the 4th Infantry Division Association publication "The Ivy Leaves" and for the Division Association's website and Facebook pages.

He is the author of the book *Steadfast And Loyal*, the stories of the Medal of Honor and Distinguished Service Cross recipients of the 4th Division in World War One. He is also the webmaster and one of the four founding members of the 1st Battalion 22nd Infantry website at www.1-22infantry.org.

In 2012, the title of Distinguished Member of the Regiment (DMOR) was conferred upon him by the Secretary of the Army and his name was added to the roll of the Distinguished Members of the Regiment of the United States 22nd Infantry.

He lives with his wife, Margaret, and their cats Bijou and Scottie in the heart of Cajun Country near Carencro, Louisiana.

www.ingramcontent.com/pod-product-compliance
Lightning Source LLC
Chambersburg PA
CBHW021922190326
41519CB00009B/885